SIMPLIFIED WASTEWATER TREATMENT PLANT OPERATIONS

WORKBOOK

Edward J. Haller, B.S. Ch.E

CRC PRESS

Boca Raton   London   New York   Washington, D.C.

Library of Congress Cataloging-in-Publication Data

Main entry under title:
Workbook—Simplified Wastewater Treatment Plant Operations

Visit the CRC Press Web site at www.crcpress.com

© 1995 by CRC Press LLC

No claim to original U.S. Government works
International Standard Book Number 1-56676-217-0
Printed in the United States of America 1 2 3 4 5 6 7 8 9 0
Printed on acid-free paper

Table of Contents

CHAPTER 1

Interceptors and Wet Well Pumping

QUESTIONS

1. What is the wastewater called that comes from the household?

2. What is the wastewater called that comes from manufacturing plants?

3. If domestic waste and industrial waste are combined, what is the wastewater called?

4. What kind of sewer contains only municipal waste?

5. What kind of sewer contains only storm water?

6. What kind of sewer contains both municipal waste and storm water?

7. If an interceptor with both municipal waste and storm water overflows into a receiving stream, what is the event called?

8. What problem could be caused by loose sewer line joints?

9. If there is an increase in the amount of sand coming into the plant, what might this be an indication of?

10. What is the ideal velocity for flow to go through an interceptor?

PROBLEMS

1. A lift station wet well is 10 ft by 12 ft. For 5 minutes the influent valve is closed and the well level drops 2.1 ft. What is the pumping rate in gallons per minute?

2. A lift station wet well is 3.1 m by 3.6 m. For 5 min the influent valve is closed and the well level drops 0.7 m. What is the pumping rate in liters per second (L/s)?

3. The influent valve to an 11 ft by 13 ft lift station wet well is closed for 4 min. During this time the well level dropped 1.8 ft. What is the pump discharge in gpm?

4. The influent valve to a 3.4 m by 4 m lift station wet well is closed for 4 min. During this time the well level dropped 0.5 m. What is the pump discharge in L/s?

5. The dimensions of the wet well for a lift station is 10 ft-9 in. by 12 ft-2 in. The influent valve to the well is closed only long enough for the level to drop 3 ft. The time to accomplish this was 6 min and 35 s. At what rate, in gallons per minute, is the pump discharging?

6. The dimensions of the wet well for a lift station is 328 cm by 372 cm. The influent valve to the well is closed only long enough for the level to drop 50 cm. The time to accomplish this was 3 min and 37 s. At what rate, in liters per second, is the pump discharging?

7. A lift station wet well is 11.5 ft by 13 ft. The influent flow to this well is 370 gpm. If the well level rises 1 in. in 7 min, how many gallons per minute is the pump discharging?

8. A lift station wet well is 3.3 m by 4.1 m. The influent flow to this well is 24 L/s. If the well level rises 3 cm in 6 min, how many liters per second is the pump discharging?

9. A lift station wet well is 130 in. by 146 in. The influent flow to this well is 450 gpm. If the well level drops 1.5 in. in 6 min, how many gallons per minute is the pump discharging?

10. A lift station wet well is 335 cm by 370 cm. The influent flow to this well is 30 L/s. If the well level drops 2.5 cm in 7 min, how many liters per second is the pump discharging?

11. A lift station wet well is 9.5 ft by 13 ft and has an influent rate of 750 gpm. The level in the well drops 7 in. in 14 min and two pumps in operation. If the first pump discharges at a rate of 490 gpm, at what pumping rate is the second pump discharging?

12. A lift station wet well is 2.9 m by 4 m and has an influent rate of 47.2 L/s. The level in the well drops 12 cm in 15 min and two pumps in operation. If the first pump discharges at a rate of 30 L/s, at what pumping rate is the second pump discharging?

Preliminary Treatment

QUESTIONS

1. What materials are often found in the screenings?

2. How many cubic feet of screenings will a plant remove from each million gallons of wastewater received?

3. How many cubic meters of screenings will a plant remove from each million cubic meters of wastewater received?

4. Name two types of coarse influent screens.

5. What are the bar spacings for each type?

6. Name three ways to operate a mechanical rake.

7. Name two devices that could be used to monitor water level.

8. What is the best velocity for the flow to go through a bar screen?

9. Below what velocity will grit settle in the screening channel?

10. What percentage of the screenings is typically water?

11. Name four things that may be found in grit.

12. Name two reasons why we would want to remove grit.

13. Name two kinds of grit removal units.

14. How slow should the wastewater be in order to settle the grit?

15. If an aerated grit channel has too much air, what happens to the grit?

16. What shape would you describe the water path in an aerated grit channel?

PROBLEMS

1. An empty screenings hopper 4.2 ft by 5.6 ft is filled to an even depth of 26 in. over the course of 96 hr. If the average plant flowrate was 4.8 MGD during this period, how many cubic feet of screenings were removed per million gallons of wastewater received?

2. An empty screenings hopper 1.23 m by 1.54 m is filled to an even depth of 95 centimeters over the course of 4 days. If the average plant flowrate was 29 ML/d during this period, how many cubic meters of screenings were removed per million cubic meters of wastewater received?

3. A grit channel has a water depth of 1.1 ft and width of 1.3 ft. The flowrate through the channel is 650 gpm. What is the velocity through the channel in ft/s?

4. A grit channel has a water depth of 9.32 m and a width of 9.41 m. The flowrate through the channel is 40 L/s. What is the velocity through the channel in m/s?

5. A grit channel has a water depth of 14 in. and a width of 16 in. The flowrate through the channel is 0.9 MGD. What is the velocity through the channel in ft/s?

6. A grit channel has a water depth of 31.5 cm and a width of 41.2 cm. The flowrate through the channel is 39.8 L/s. What is the velocity through the channel in m/s?

CHAPTER 3

Plant Loadings

QUESTIONS

1. What is the ceramic filter holder used in the suspended solids test called?

2. At what temperature is the filter dried?

3. What test is used to measure biological activity?

4. What is the volume of a BOD bottle?

5. What happens in a BOD test that uses oxygen?

6. How long are BOD samples incubated?

7. Generally speaking, how much DO can a BOD sample start with?

PROBLEMS

1. A suspended solids test was done on a 50 mL sample. The weight of the crucible and filter before the test was 25.6732 g. After the sample was filtered and dried, the cooled crucible weight was 25.6829 g. What was the concentration of suspended solids in mg/L?

2. A 26.1349 g crucible was used to filter 25 mL of raw influent sample for a suspended solids test. The dried crucible weighed 26.1425 g. What was the concentration of suspended solids in mg/L?

3. A BOD test was done on a 5 mL sample. The initial DO of the sample and dilution water was 7.82 mg/L. The DO of the sample after 5 days of incubation was 4.17 mg/L. What was the BOD of the sample?

4. A BOD test was done on a 5 mL sample. The initial DO of the sample and dilution water was 7.84 mg/L. The DO of the sample after 5 days of incubation was 4.08 mg/L. What was the BOD of the sample?

5. A wastewater treatment plant receives a flow of 2.74 MGD with a total phosphorus concentration of 14.3 mg/L. How many pounds per day?

6. A treatment plant with an influent flowrate of 9350 m³/d has an influent total phosphorus concentration of 14.3 mg/L. How many kg/d?

,

7. Raw influent BOD is 290 mg/L. If the influent flowrate is 5.70 MGD, at what rate are the pounds of BOD entering the plant?

8. Raw influent BOD is 290 mg/L. If the influent flowrate is 19,500 m^3/d, at what rate are the kilograms of BOD entering the plant?

9. The plant's influent flowrate of 4.65 MGD has a suspended solids concentration of 192 mg/L. How many pounds of suspended solids enter daily?

10. The plant's influent flowrate of 17,600 m³/d has a suspended solids concentration of 192 mg/L. How many kilograms of suspended solids enter daily?

CHAPTER 4

Primary Treatment

QUESTIONS

1. What is the main purpose of a primary clarifier?

2. How are settleable solids measured?

3. How much of the settleable solids are removed by primary settling?

4. How is total solids concentration measured?

5. What percentage of the total solids are removed by primary settling?

6. What is the BOD removal rate in a PST?

7. What percentage of the suspended solids are removed in the primaries?

8. What is an average detention time in a primary clarifier?

9. What two chemicals do wastewater treatment plants try to reduce in concentration?

10. What process follows primary treatment?

PROBLEMS

1. The influent flowrate to a PST is 1.86 MGD. The tank is 82 ft in length, 20 ft wide and has a water depth of 12.5 ft. What is the detention time of the tank in hours?

2. The influent flowrate to a PST is 7041 m^3/d. The tank is 25 m in length, 6.1 m wide and has a water depth of 3.8 m. What is the detention time of the tank in hours?

3. A PST 80 ft long, 18 ft wide and 12 ft deep receives a flowrate of 1.35 MGD. What is the surface overflow rate in gpd/ft^2?

4. A PST 23 m long, 5.4 m wide and 3.6 m deep receives a flowrate of 5062 m³/d. What is the surface overflow rate in m³/d/m²?

5. A primary sludge sample is tested for total solids. The dish alone weighed 21.34 g. The sample with the dish weighed 74.56 g. After drying, the dish with dry solids weighed 22.07 g. What was the percent total solids (%TS) of the sample?

6. Primary sludge is pumped to a gravity thickener at 390 gpm. The sludge concentration is 0.9%. How many pounds of solids are pumped daily?

7. Primary sludge is pumped to a gravity thickener at 24.6 L/s. The sludge concentration is 0.9%. How many kilograms of solids are pumped daily?

8. The raw influent suspended solids concentration is 135 mg/L. The primary effluent concentration of suspended solids is 52 mg/L. What percentage of the suspended solids is removed by primary treatment?

9. The primary influent settleable solids are 14.5 mL/L. The settleable solids of the primary effluent are 1.3 mL/L. What percentage of the settleable solids does the primary treatment remove?

10. A primary tank with a total weir length of 70 ft receives a flowrate of 1.38 MGD. What is the weir overflow rate in gpd/ft?

11. A primary tank with a total weir length of 21.5 m receives a flowrate of 5175 m³/d. What is the weir overflow rate in m³/d/m?

CHALLENGE PROBLEMS

12. A wastewater treatment plant has eight primary tanks. Each tank is 82 ft long, 21 ft wide with a side water depth of 13 ft and a total weir length of 88 ft. The flowrate to the plant is 5.1 MGD. There are three tanks currently in service.

    a. Calculate the detention time in minutes, the surface overflow rate (SOR) in gpd/ft² and the weir overflow rate (WOR) in gpd/ft.

    b. Plant flowrate has increased to 9.8 MGD because of storm flow. How many of the eight PSTs should be in service to operate with these conditions:

        1. Detention time between 90 and 150 min
        2. SOR between 800 and 1000 gpd/ft²
        3. WOR below 20,000 gpd/ft

13. A wastewater treatment plant has eight primary tanks. Each tank is 25 m long, 6.4 m wide with a side water depth of 4 m and a total weir length of 26.8 m. The flowrate to the plant is 19,300 m³/d. There are three tanks currently in service.

 a. Calculate the detention time in minutes, the surface overflow rate (SOR) in m³/d/m² and the weir overflow rate (WOR) in m³/d/m.

b. Plant flowrate has increased to 37,100 m^3/d because of storm flow. How many of the eight PSTs should be in service to operate with these conditions:

1. Detention time between 90 and 150 min
2. SOR between 32 and 41 m^3/d/m^2
3. WOR below 250 m^3/d/m

CHAPTER 5

Trickling Filters

QUESTIONS

1. What is the slime called that grows in a trickling filter?

2. What does a trickling filter filter?

3. What is the process called in which the slime layer gets too thick and falls off?

4. Is the hydraulic loading rate based on the entire volume of the trickling filter or just the top surface?

5. What is the recirculation ratio?

6. Describe one way to control the recirculation rate.

7. Why would we want to recirculate unclarified effluent?

8. Is the organic loading rate based on the entire volume of the trickling filter or just the top surface?

9. Name two kinds of synthetic media.

10. What percentage of rock media is open space?

11. What percentage of synthetic media is open space?

12. Why do we need open space in a trickling filter?

13. Give another name for a trickling filter with a media bed depth greater than 10 ft.

14. Name four categories of trickling filters based on their organic loading rate.

15. What hydraulic loading rate should be used with the high rate plastic media?

16. What is ponding?

17. During winter operation, what happens to the water temperature as the recirculation is increased?

18. What role might the worms and snails play in the operation of a trickling filter?

19. If bed temperatures get too cold, what happens to the worms and snails?

20. How can an operator determine if the water is being evenly distributed over the filter?

21. What can be used to control filamentous organism growth in a trickling filter?

PROBLEMS

1. A trickling filter 80 ft in diameter treats a primary effluent flowrate of 0.296 MGD. If the recirculated flow to the clarifier is 0.348 MGD, what is the hydraulic loading rate on the trickling filter in gallons per day per square foot (gpd/ft²).

2. A trickling filter 25 m in diameter treats a primary effluent flowrate of 1046 m^3/d. If the recirculated flow to the clarifier is 1130 m^3/d, what is the hydraulic loading rate on the trickling filter in cubic meters per day per square meter media surface (m^3/d/m^2)?

3. A treatment plant receives a flowrate of 2.5 MGD. If the trickling filter effluent is recirculated at a rate of 4.25 MGD, what is the recirculation ratio?

4. A treatment plant receives a flowrate of 9500 m³/d. If the trickling filter effluent is recirculated at a rate of 16,000 m³/d, what is the recirculation ratio?

5. A trickling filter 70 ft in diameter with a media depth of 5 ft receives a primary effluent flowrate of 1,150,000 gpd. If the PE BOD is 74 mg/L, what is the organic loading rate on the unit in pounds per day per 1000 cubic feet (lbs/d/1000 ft³)? What is the loading rate in pounds per day per ac-ft (lbs/d/ac-ft)?

6. A trickling filter 21 m in diameter with a media depth of 1.5 m receives a primary effluent flowrate of 4350 m³/d. If the PE BOD is 74 mg/L, what is the organic loading rate on the unit in kilograms per day per cubic meter (kg/d/m³)?

7. The flowrate to a trickling filter is 3.7 MGD. If the PE BOD is 79 mg/L and the SE BOD is 14 mg/L, how many pounds of BOD are removed daily?

8. The flowrate to a trickling filter is 14,000 m³/d. If the PE BOD is 79 mg/L and the SE BOD is 14 mg/L, how many kilograms of BOD are removed daily.

CHALLENGE PROBLEMS

9. The organic loading rate applied to a high-rate plastic media filter is 24 lb BOD/d per 1000 ft³. The unit is 25 ft in diameter and 12 ft deep. If the primary effluent applied has a BOD concentration of 80 mg/L, at what rate in gpm should the clarified effluent be recirculated?

10. The organic loading rate applied to a high-rate plastic media filter is 39 kg BOD/d per 100 m³. The unit is 7.6 m in diameter and 3.6 m deep. If the primary effluent applied has a BOD concentration of 80 mg/L, at what rate in L/s should the clarified effluent be recirculated?

11. A wastewater treatment plant has three trickling filters. Each trickling filter is 20 m in diameter and has a media depth of 5 ft. The present hydraulic loading rate is 0.3 L/s/m². The filters experience a 92% BOD removal efficiency. Both of the filters in service experience an organic loading rate between 40 lb BOD/d/1000 ft³ and 160 kg BOD/d/100 m³. If the present recirculated flowrate is 5000 m³/d, by how many cubic meters per day should the recirculated flow be decreased in order to lower the hydraulic loading rate to 0.25 L/s/m²?

CHAPTER 6

Rotating Biological Contactors

QUESTIONS

1. What is the purpose of an RBC?

2. Can an RBC be operated without primary settling?

3. What percentage of an RBC disk is submerged?

4. How large are the RBC media disks?

5. How long is an RBC shaft?

6. How much surface area does a standard media shaft provide?

7. In what range of rotational speeds are RBCs typically operated?

8. Why should an RBC be followed by a settling tank?

9. Give four reasons to keep an RBC covered.

10. What is an RBC stage?

11. Describe a health first stage biomass.

12. What does white biomass indicate?

13. What do filamentous bacteria eat?

14. What chemical could be added to increase the dissolved oxygen (DO) concentration?

15. What is the minimum DO that should be leaving the first stage of an RBC?

16. What is the minimum DO that should be leaving the last stage of an RBC?

17. What range of hydraulic loading is generally applied to an RBC?

18. Describe the biomass of an organically overloaded first stage.

19. If soluble BOD is one part of the total BOD, what is the other part?

20. What change is made to the total BOD test in order to make the test for soluble BOD?

21. How can the particulate BOD concentration be calculated from the suspended solids concentration?

22. Domestic sewage typically has what value of k?

23. What range of organic loading does an RBC typically receive?

24. What range of organic loading does just the first stage of an RBC typically receive?

25. Name two types of RBC drive units.

26. Name three advantages of air drives.

27. Name three disadvantages of air drives.

28. What is RBC media made of?

29. How much surface area does a standard density media shaft provide?

30. Is the third stage biomass thicker or thinner than the first stage biomass?

31. Name two types of RBC media.

32. Why does high density media have more surface area available?

33. Why wouldn't high density media be installed in the first stage?

34. How much surface area is available on each shaft of high density media?

35. What do snails do that cause a problem in the final RBC stage?

36. Is the period from the time snail eggs are laid to the time they hatch dependent on the wastewater temperature?

37. What chemical has been effectively used in three doses to cure a snail problem?

PROBLEMS

1. An RBC unit treats a flowrate of 0.35 MGD. The two shafts used provide a total surface area of 195,000 ft^2. What is the hydraulic loading on the unit in gpd/ft^2?

2. An RBC unit treats a flowrate of 1320 m^3/d. The two shafts used provide a total surface area of 18,100 m^2. What is the hydraulic loading on the unit in m^3/d/m^2?

3. The influent to an RBC has a total BOD concentration of 180 mg/L and a suspended solids concentration of 200 mg/L. If there are 0.5 pounds of particulate BOD per pound of suspended solids, estimate the soluble BOD concentration in mg/L.

4. An RBC receives a flowrate of 1.4 MGD. If the influent soluble BOD concentration is 122 mg/L and the total media surface area is 392,000 ft² for the RBC unit, what is the organic loading in lbs/d/1000ft²?

5. An RBC receives a flowrate of 5300 m^3/d. If the influent soluble BOD concentration is 122 mg/L and the total media surface area is 36,400 m^2 for the RBC unit, what is the organic loading in kg/d/100 m^2?

6. Estimate the soluble BOD loading on an RBC treating a flowrate of 0.69 MGD. The total unit surface area is 104,000 ft^2. The total BOD concentration is 202 mg/L with a suspended solids concentration of 225 mg/L and a k value of 0.65.

7. Estimate the soluble BOD loading on an RBC treating a flowrate of 2610 m³/d. The total unit surface area is 9660 m². The total BOD concentration is 202 mg/L with a suspended solids concentration of 225 mg/L and a k value of 0.65.

CHALLENGE PROBLEMS

8. An RBC unit contains two shafts operated in series each with a surface area of 102,000 ft². The shafts can both be partitioned by baffles at 25% shaft length intervals. Currently the first stage of the RBC unit is baffled to use 75% of one of the two shafts. The unit receives a flowrate of 0.445 MGD. The primary effluent total BOD concentration is 237 mg/L. The suspended solids concentration is 148 mg/L and the value of k is 0.5. Calculate:
   a) Hydraulic loading

b) Unit organic loading

c) First stage organic loading

Should anything be changed?

9. An RBC unit contains two shafts operated in series with a surface area of 9475 m². The shafts can both be partitioned by baffles at 25% shaft length intervals. Currently the first stage of the RBC unit is baffled to use 75% of one of the two shafts. The unit receives a flowrate of 1714 m³/d. The primary effluent total BOD concentration is 237 mg/L. The suspended solids concentration is 148 mg/L and the value of k is 0.5. Calculate:
   a) Hydraulic loading

b) Unit organic loading

c) First stage organic loading

Should anything be changed?

CHAPTER 7

Activated Sludge

QUESTIONS

1. What two things does the air being put into an aeration tank provide?

2. What is the watery mixture of bugs and solids removed from the bottom of the settling tank called?

3. Primary effluent and return sludge are combined in the aeration tank. Which of these contains the "bugs" and which contains the "food"?

4. What is the mixture of primary effluent and return sludge called?

5. The activated sludge removed from the bottom of the clarifier is split into two different flows. What are each of these called?

6. What is the major purpose for an activated sludge unit?

7. What is the official name for the little bugs?

8. What kind of little bugs eat things like spaghetti sauce and other carbon-containing foods?

9. What kind of little bugs eat nitrogen-containing chemicals?

10. How do the small bugs stick together?

11. What big bug might be called the "blob"?

12. Which of the big bugs uses a whip to move around?

13. What is the first type of "hairy" big bug to show up after a new plant start-up?

14. What two kinds of big bugs does the operator hope to have in greatest number?

15. Does an operator want filaments in the sludge?

16. What will happen to the sludge if too many filaments are in it?

17. What flow is always put first into an aeration tank?

18. Describe a plug-flow aeration tank.

19. Name one advantage and one disadvantage to using a plug-flow aeration tank.

20. What is the difference between step aeration and step feed?

21. How might the air distribution be set up in a four-pass tapered aeration tank?

22. How much contact time should be allowed for the bugs to get their food?

23. How much stabilization time do the bugs need to digest their food?

24. Name some advantages to using contact stabilization.

25. How can step feed reduce oxygen requirements?

26. How can changing from big-bubble to small-bubble diffusers reduce air requirements?

27. What are the three main things an operator can control when operating an activated sludge unit?

28. What kind of bugs might grow in larger numbers if there isn't enough air put into the aeration tank?

29. What is a safe minimum dissolved oxygen concentration to stay above when operating an activated sludge unit?

30. What is the name given to the sludge settled on the bottom of an FST?

31. How can an operator measure the depth of the sludge blanket in a clarifier?

32. What's a good sludge blanket thickness to keep?

33. What kind of test shows the operator how the sludge is settling in the clarifier?

34. What size settlometer is best to use? Why?

35. What is the disadvantage to using a constant return pumping rate?

36. Under what plant conditions might a target MLSS be fine to use?

37. What is the maximum percentage the wasting rate should be changed in one day?

38. How would the target MLSS concentration be changed for summer operation?

39. What is the difference between MCRT and SRT?

40. Is the sludge age based on the bugs coming into or those leaving the activated sludge unit?

41. What two paths can the bugs take to leave an activated sludge unit?

42. Should the bugs in the return sludge be used in an SRT calculation?

43. If an operator has already measured the suspended solids concentration of a sample, how can the volatile suspended solids concentration be measured?

44. BOD test takes five days to complete. How can the operator get *F/M* data in less time?

45. How can the operator use the *F/M* data to control wasting rates if the *F/M* data makes frequent large but temporary changes?

46. If you want to look at big bugs under the microscope, what magnification should be used?

47. If the bugs are moving around too fast to get a good look at them, how can they be slowed down?

48. Which two of the big bugs would be best to have in greatest number in the sludge?

49. What method makes use of a lab centrifuge?

50. If nocardia bugs are causing a foam problem, how many more of them will be in the foam compared to those in the mixed liquor?

51. Do nocardia bugs cause problems with sludge settling?

52. Name two properties that nocardia bugs have which allow them to be such a problem in the foam.

53. Describe plant conditions that would favor breeding of nocardia bugs.

54. How do the dead nocardia bugs add to the problem?

55. How might nocardia foam be removed from the aeration tanks?

56. How would you describe bulking sludge?

57. To get an early indication of potential sludge bulking, how often should the sludge be examined by microscope?

58. If the return sludge is chlorinated, how much chlorine should be dosed?

59. If chlorination of the return sludge is used, how long is it before the sludge settleability should improve?

60. What are some temporary remedies for a sludge bulking problem?

PROBLEMS

1. A mixed liquor sample is poured into a 2000 mL settlometer. After 30 min there is a settled sludge volume of 430 mL. If the plant flowrate (Q) is 5.9 MGD, what should the return sludge flowrate be in gpm?

2. A mixed liquor sample is poured into a 2000 mL settlometer. After 30 min there is a settled sludge volume of 430 mL. If the plant flowrate (Q) is 22,300 m³/d, what should the return sludge flowrate be in m³/d?

3. The mixed liquor in a 0.42 MG Aeration Tank has a suspended solids concentration (MLSS) of 2200 mg/L. The waste sludge is being removed at a rate of 0.118 MGD and has a concentration of 4950 mg/L. If the target MLSS is 2150 mg/L, what should the new waste sludge pumping rate be?

4. The mixed liquor in a 1590 m³ Aeration Tank has a suspended solids concentration (MLSS) of 2200 mg/L. The waste sludge is being removed at a rate of 447 m³/d and has a concentration of 4950 mg/L. If the target MLSS is 2150 mg/L, what should the new waste sludge pumping rate be?

5. The mixed liquor in a 0.42 MG Aeration Tank has a suspended solids concentration (MLSS) of 2080 mg/L. The waste sludge is being removed at a rate of 86.1 gpm and has a concentration of 4890 mg/L. If the target MLSS is 2150 mg/L, what should the new waste sludge pumping rate be in gpm?

6. The mixed liquor in a 1590 m³ Aeration Tank has a suspended solids concentration (MLSS) of 2080 mg/L. The waste sludge is being removed at a rate of 5.43 L/s and has a concentration of 4890 mg/L. If the target MLSS is 2150 mg/L, what should the new waste sludge pumping rate be in L/s?

7. An aeration tank has an MLSS concentration of 2150 mg/L. The volume of the tank is 0.54 MG. The plant flowrate is 3.14 MGD. The primary effluent suspended solids concentration is 128 mg/L. What is the sludge age?

8. An aeration tank has an MLSS concentration of 2150 mg/L. The volume of the tank is 2040 m³. The plant flowrate is 11,890 m³/d. The primary effluent suspended solids concentration is 128 mg/L. What is the sludge age?

9. An aeration tank has an MLSS concentration of 2180 mg/L. The volume of the tank is 0.48 MG. The plant flowrate is 2.92 MGD. The waste sludge suspended solids concentration is 5340 mg/L. This sludge is being removed at a rate of 0.095 MGD. The secondary effluent suspended solids concentration is 17 mg/L. What is the Solids Retention Time?

10. An aeration tank has an MLSS concentration of 2180 mg/L. The volume of the tank is 1820 m³. The plant flowrate is 11,050 m³/d. The waste sludge suspended solids concentration is 5340 mg/L. This sludge is being removed at a rate of 360 m³/d. The secondary effluent suspended solids concentration is 17 mg/L. What is the Solids Retention Time?

11. Calculate the SRT from the following data.

Aer. Vol. = 3.08 MG	MLSS = 2040 mg/L
FST Vol. = 0.23 MG	WAS SS = 4980 mg/L
Plant Flow = 6.82 MGD	S.E. SS = 16 mg/L
Waste Rate = 0.154 MGD	

Clarifier Core Sample SS = 1500 mg/L

In this calculation include the bugs in the clarifier. This is done by taking a core sample of the clarifier (FST) and running a suspended solids test on the sample. This concentration is then used as the uniform concentration in the whole clarifier.

12. Calculate the SRT from the following data.

Aer. Vol. $= 11,660$ m³   MLSS $= 2040$ mg/L
FST Vol. $= 870$ m³   WAS SS $= 4980$ mg/L
Plant Flow $= 25,820$ m³/d   S.E. SS $= 16$ mg/L
Waste Rate $= 583$ m³/d

Clarifier Core Sample SS $= 1500$ mg/L

In this calculation include the bugs in the clarifier. This is done by taking a core sample of the clarifier (FST) and running a suspended solids test on the sample. This concentration is then used as the uniform concentration in the whole clarifier.

13. Use the target Solids Retention Time to determine what the wasting rate should be in gpm.

   Aer. Tank Vol. = 0.56 MG   WAS SS = 5120 mg/L
   Plant Flowrate = 3.44 MGD   MLSS = 1980 mg/L
   Target SRT = 2.95 days   S.E. SS = 15 mg/L

14. Use the target Solids Retention Time to determine what the wasting rate should be in L/s.

   Aer. Tank Vol. = 2120 m^3 WAS SS = 5120 mg/L
   Plant Flowrate = 13,020 m^3/d   MLSS = 1980 mg/L
   Target SRT = 2.95 days   S.E. SS = 15 mg/L

15. Using the core sample method, calculate the waste sludge pumping rate from the target SRT using the following data.

   Aer Vol. = 3.08 MG   MLSS = 2131 mg/L
   FST Vol. = 0.23 MG   WAS SS = 5060 mg/L
   Plant Flow = 6.78 MGD   S.E. SS = 14 mg/L
   Target SRT = 7.4 days

   Clarifier Core Sample SS = 1710 mg/L

16. Using the core sample method, calculate the waste sludge pumping rate from the target SRT using the following data.

Aer. Vol. = 11,660 m³   MLSS = 2131 mg/L
FST Vol. = 870 m³   WAS SS = 5060 mg/L
Plant Flow = 25,660 m³/d   S.E. SS = 14 mg/L
Target SRT = 7.4 days

Clarifier Core Sample SS = 1710 mg/L

17. An aeration tank with a volume of 0.48 MG receives a primary effluent flowrate of 2.56 MGD. The mixed liquor suspended solids concentration is 2230 mg/L. If the primary effluent BOD concentration is 172 mg/L, what is the current F/M ratio?

18. An aeration tank with a volume of 1820 m³ receives a primary effluent flowrate of 9690 m³/d. The mixed liquor suspended solids concentration is 2230 mg/L. If the primary effluent BOD concentration is 172 mg/L, what is the current F/M ratio?

19. A 25 mL of mixed liquor is tested for suspended solids and then for volatile suspended solids. Given the following information, calculate the concentration of suspended solids in mg/L, the percent volatile suspended solids, and the concentration of volatile suspended solids in mg/L.

	After Drying	After Burning
Weight of Sample + Crucible	23.4791 g	23.4406 g
Weight of Crucible	23.4256 g	23.4256 g

20. The following data is the primary effluent COD and BOD concentrations from earlier this month. The current COD concentration of the primary effluent is 180 mg/L. Estimate the current concentration of BOD in the primary effluent.

Date	P.E. COD	P.E. BOD
6-1	174 mg/L	106 mg/L
6-2	157 mg/L	80 mg/L
6-3	211 mg/L	99 mg/L
6-4	186 mg/L	110 mg/L
6-5	163 mg/L	88 mg/L

21. An aeration tank with a volume of 0.48 MG receives a primary effluent flowrate of 2.64 MGD. The mixed liquor volatile suspended solids are 1940 mg/L. The primary effluent COD concentration is 184 mg/L. The current BOD/COD factor is 0.58. Calculate the *F/M* ratio using the MLVSS and the estimated current BOD concentration.

22. An aeration tank with a volume of 1820 m³ receives a primary effluent flowrate of 10,1000 m³/d. The mixed liquor volatile suspended solids are 1940 mg/L. the primary effluent COD concentration is 184 mg/L. The current BOD/COD factor is 0.58. Calculate the *F/M* ratio using the MLVSS and the estimated current BOD concentration.

23. Use the target Food to Microorganism Ratio to determine the new sludge wasting rate using the following data:

Aer. Tank Vol. = 0.58 MG   WAS SS = 2090 mg/L
Plant Flowrate = 3.18 MGD   MLSS = 2090 mg/L
Wasting Rate = 0.122 MGD   PE BOD = 158 mg/L
Target F/M = 0.42

24. Use the target Food to Microorganism Ratio to determine the new sludge wasting rate using the following data:

    Aer. Tank Vol. = 2200 m³   WAS SS = 5150 mg/L
    Plant Flowrate = 12,040 m³/d   MLSS = 2090 mg/L
    Wasting Rate = 462 m³/d   PE BOD = 158 mg/L
    Target F/M = 0.42

CHALLENGE PROBLEMS

25. An aeration tank has a volume of 1.38 MG. The final clarifier is 0.117 MG. The secondary effluent suspended solids concentration is 20 mg/L. The primary effluent flowrate is 2.9 MGD. Sludge is wasted at a rate of 75,000 gpd. MLSS concentration is 2560 mg/L. The WAS concentration is 5960 mg/L. A core sample of the clarifier has a suspended solids concentration of 1900 mg/L. The aeration tanks are covered with a troublesome foam. Water sprays and antifoamant chemicals do not help.

a. Using the core sampler method, determine the SRT.

b. What can be done to solve the foam problem?

26. An aeration tank has a volume of 5220 m³. The final clarifier is 442 m³. The secondary effluent suspended solids concentration is 20 mg/L. The primary effluent flow is 10,980 m³/d. Sludge is wasted at a rate of 284 m³/d. MLSS concentration is 2560 mg/L. The WAS concentration is 5960 mg/L. A core sample of the clarifier has a suspended solids concentration of 1900 mg/L. The aeration tanks are covered with a troublesome foam. Water sprays and antifoamant chemicals do not help.

a. Using the core sampler method, determine the SRT.

b. What can be done to solve the foam problem?

CHAPTER 8

Nitrification and Denitrification

QUESTIONS

1. Organic material can be thought of as long chains of what kind of things?

2. What is an ion?

3. What kind of charge does an ammonium ion have?

4. Describe an ammonium ion.

5. What form of nitrogen is often highest in concentration in the influent of most municipal wastewater treatment plants?

6. What are the two different forms of ammonia?

7. Can these forms change back and forth?

8. At a pH of 7.0, what form of ammonia is mostly present?

9. Which nitrogen bug eats ammonia?

10. Which nitrogen bug eats nitrite?

11. What is the process called that changes ammonia to nitrate?

12. What is the "Total Potential Ammonia"?

13. What term could be used to describe a tank condition with no dissolved oxygen present but with some nitrate present?

14. Give two examples of chemicals containing chemically combined oxygen.

15. What is denitrification?

16. Why is denitrification not desirable in a clarifier?

17. Are the nitrifiers younger, the same age, or older than the carbon eating small bugs?

18. Where do the carbon eating bugs come from?

19. Where do the nitrifiers come from?

20. What is alkalinity?

21. How is alkalinity expressed?

22. How much alkalinity is needed for every pound of ammonia changed to nitrate?

23. How will the effluent look if there are too many nitrifiers and not enough of the carbon eaters in the mixed liquor?

24. What BOD:TKN ratio is it best to operate above?

25. How can the BOD:TKN ratio be raised?

CHAPTER 9

Biological Nutrient Removal

QUESTIONS

1. What kind of bugs require oxygen to live?

2. What kind of bugs live strictly without oxygen?

3. What kind of bugs can live with or without oxygen?

4. Which of these types of bugs removes the most amount of phosphorus?

5. What should the operator provide the facultative bugs with under anaerobic conditions to give them a bigger advantage over the aerobic bugs?

6. Name two categories of biological phosphorus removal processes.

7. What change in plant operation are some newer facilities doing to help with biological phosphorus removal?

8. How can nitrate in the return sludge cause problems in the anaerobic zone?

9. When a plant is using a main stream biological phosphorus removal process what concern might be raised about the solids handling recycles?

10. Under what conditions does it require more energy for facultative bugs to eat?

11. What detention time is typically used in a stripper?

12. What is the difference between a return flow and a recirculated flow?

CHALLENGE PROBLEM

A dual nutrient removal process treats an average daily flowrate of 5.7 MGD or 21,580 m³/d. After preliminary treatment, return sludge is added to the wastewater. The mixed liquor then proceeds through five different tanks before the mixed liquor is finally settled in a clarifier. The first tank has anaerobic conditions to promote biological phosphorus removal. Aeration mixed liquor is recirculated ahead of the second tank to convert nitrate to nitrogen gas under anoxic conditions. The mixed liquor is then aerated for the reduction of BOD as well as for nitrification. The fourth tank provides a second anoxic contact to promote denitrification. Finally, the mixed liquor is reaerated in the fifth tank before clarification. Waste pickle liquor in the form of ferrous sulfate is added prior to reaeration to reduce effluent phosphorus concentrations.

Design Information

Tank Type	Size	Minimum Detention Time
Anaerobic	0.8 MG or 3028 m³	2 hours
Anoxic	2.0 MG or 7571 m³	3 hours
Aerobic	8.0 MG or 30,280 m³	12 hours
2nd Anoxic	0.8 MG or 3028 m³	2 hours
Reaeration	0.4 MG or 1514 m³	1 hour

Anoxic Zone D.O. < 0.2 mg/L

Effluent Limits: TP – 1.0 mg/L
NH$_3$ – 1.0 mg/L
TN – 5.0 mg/L

Current Data

Effluent Concentrations: TP – 0.82 mg/L
NH$_3$ – 0.70 mg/L
NO$_2$ – 0.09 mg/L
NO$_3$ – 7.42 mg/L
TKN – 2.46 mg/L

Return Sludge Flowrate = 2.8 MGD or 10,600 m³/d
Recirculated Mixed Liquor Flowrate = 2.2 MGD or 8330 m³/d

a. Calculate the hydraulic detention time of the first anoxic zone.

b. What should be changed to improve effluent quality?

CHAPTER 10

Waste Treatment Ponds

QUESTIONS

1. Name three categories of lagoons.

2. What type of pond is the most common?

3. What are the two most common sources of oxygen for lagoons?

4. What problem with effluent quality can the use of algae cause?

5. What solutions could be considered to solve this algae problem?

6. How deep is a facultative lagoon usually built?

7. A lack of wave action on the lagoon surface might indicate what?

8. When carbon dioxide is added to water, how is the pH affected?

9. During a twenty-four hour cycle in the summer, how does the algae affect the dissolved oxygen concentration and the pH in a lagoon?

10. What does a dark green lagoon color indicate?

11. What causes a dull green to yellow color to develop?

12. How can muskrats be discouraged from burrowing around a lagoon?

13. What chemical can be added to a lagoon to increase available oxygen?

14. What is the purpose of an anaerobic lagoon?

15. How can the odors from an anaerobic lagoon be reduced if the unit has just recently been started up?

PROBLEMS

1. A wastewater treatment pond has an average length of 705 ft with an average width of 450 ft. If the flowrate to the pond is 290,000 gal each

day, and is operated at a depth of 5.8 ft, what is the hydraulic detention time in days?

2. A wastewater treatment pond has an average length of 215 m with an average width of 135 m. If the flowrate to the pond is 1100 m³/d, and is operated at a depth of 1.8 m, what is the hydraulic detention time in days?

3. What is the detention time for a pond receiving an influent flowrate of 0.45 ac-ft each day? The pond has an average length of 695 ft and an average width of 425 ft. The operating depth of the pond is 49 in.

4. What is the hydraulic overflow rate for the pond described in problem #3 in units of inches per day?

5. What is the hydraulic overflow rate for the pond described in problem #2 in units of centimeters per day?

6. A waste treatment pond has an average width of 387 ft and an average length of 692 ft. The influent flowrate to the pond is 0.14 MGD with a BOD concentration of 162 mg/L. What is the organic loading rate to the pond in pounds per day per acre?

7. A waste treatment pond has an average width of 118 m and an average length of 206 m. The influent flowrate to the pond is 559 m³/d with a BOD concentration of 162 mg/L. What is the organic loading rate to the pond in kilograms per day per hectare?

8. A pond 742 ft long and 427 ft wide receives an influent flowrate of 0.62 ac-ft/d. What is the hydraulic loading rate on the pond in inches per day?

9. A pond 227 m long and 124 m wide receives an influent flowrate of 792 m³/d. What is the hydraulic loading rate on the pond in centimeters per day?

CHALLENGE PROBLEMS

10. A facultative pond has been recently experiencing nuisance odors
    which have generated complaints from the community. It has been
    decided to apply sodium nitrate to the pond. Recommended application
    is 100 lb/ac the first day and 50 lb/ac each day thereafter until the odor
    is gone. The chemical is to be applied to the pond in the wake of a
    motor boat. The treatment pond is 450 ft by 670 ft and is 4.7 ft deep.
    a. How many pounds of sodium nitrate will be needed for the first
        seven days?

    b. What else can be done to reduce odor?

11. A facultative pond has been recently experiencing nuisance odors which have generated complaints from the community. It has been decided to apply sodium nitrate to the pond. Recommended application is 112 kg/ha the first day and 56 kg/ha each day thereafter until the odor is gone. The chemical is to be applied to the pond in the wake of a motor boat. The treatment pond is 137 m by 204 m and is 1.4 m deep.

    a. How many kilograms of sodium nitrate will be needed for the first seven days?

    b. What else can be done to reduce odor?

CHAPTER 11

Physical and Chemical Treatment

QUESTIONS

1. Name four chemicals that can be added to wastewater to remove phosphorus.

2. What is the name of the unit used to mix lime with water?

3. What effect will alum have on sludge dewatering?

4. What was pickle liquor before it was green?

5. Name two categories of filters.

6. Which category of filter is better for wastewater treatment?

7. How would you describe the media size distribution in a multiple media filter?

8. Where do the solids collect in a single media filter?

9. Where do the solids collect in a multiple media filter?

10. In what direction does the flow pass through the filter during normal operation?

11. Name two parameters used to monitor filter operation.

12. Where should polymer be added?

13. If the polymer dosage is too high, what happens to the filter operation?

14. If the polymer dosage is too low, what happens to the filter operation?

15. As water temperature rises, what happens to the required backwash flowrate?

16. As the temperature goes up, what happens to the amount of polymer needed?

17. What percentage of backwash should not be exceeded?

PROBLEMS

1. There are six effluent filters in service. Each filter treats an average flowrate of 2730 gpm. If each filter is backwashed for 9 min daily and the backwash flowrate is 8420 gpm per filter, what is the percent backwash?

2. There are six effluent filters in service. Each filter treats an average flowrate of 172 L/s. If each filter is backwashed for 9 min daily and the backwash flowrate is 530 L/s per filter, what is the percent backwash?

3. A filter is operated for 67 h. During that time a total of 14.2 MG of secondary effluent was filtered. The filter is 34 ft long and 22 ft wide with a filter bed depth of 30 in. What is the average filtration rate in gpm/ft²?

4. A filter is operated 67 hr. During that time a total of 53,750 m³ of secondary effluent was filtered. The filter is 10.4 m long and 6.7 m wide with a filter bed depth of 76 cm. What is the average filtration rate in m³/min/m²?

CHALLENGE PROBLEM

5. A dual media filter was placed in operation last February at a rural Michigan wastewater treatment facility. In the last few months, the filters have required more frequent backwashing because of effluent turbidity. Based on the data provided, determine:

    a. What is the present backwash flowrate in gpm or L/s.

    b. What should be changed?

    Operational Data

    - dimensions: 15 ft × 18 ft (4.6 m × 5.5 m)
    - media depth: 62 in. (158 cm)
    - sand depth: 18 in. (46 cm)
    - backwash rise rate: 24.2 in./min (61.5 cm/min)
    - present backwash bed suspension: 8 in. (20 cm.)

	Feb. 1993	July 1993
Filtration Influent Suspended Solids:	16 mg/L	15 mg/L
Filtration Influent Polymer Dosage:	0.085 mg/L	0.087 mg/L
Filtration Influent Temperature:	20°C	29°C

CHAPTER 12

Chlorination and Dechlorination

QUESTIONS

1. What are the bugs that are disease carriers called?

2. What is the process of killing most of the bugs called?

3. What is the amount of chlorine added called?

4. What is the amount of chlorine that gets used up called?

5. What is the amount of chlorine that is left in the water after the demand has been satisfied called?

6. Name a form of chlorine used which could also be called a strong bleach.

7. What color is chlorine gas?

8. Where should chlorine vents be placed?

9. What is the chemical symbol for free available chlorine?

10. What does chlorine form with ammonia?

11. How do we measure the amount of disease carrying bugs in the water?

12. How much chlorine should be added to the water?

13. Name two chemicals used to dechlorinate.

14. What type of instrument could be used to measure chlorine residual concentrations accurately to three decimal places?

PROBLEMS

1. The chlorine dosage for a treatment plant effluent is 8.9 mg/L. The chlorine residual is measured after the chlorine contact tanks and found to be 0.4 mg/L. What is the chlorine demand expressed in mg/L?

2. The chlorine demand for a secondary effluent is 7.8 mg/L. If a 0.3 mg/L residual is desired, how many pounds of chlorine should be dosed to a flowrate of 5.7 MGD?

3. The chlorine demand for a secondary effluent is 7.8 mg/L. If a 0.3 mg/L residual is resired, how many kilograms of chlorine should be dosed to a flowrate of 21.6 ML/d?

CHALLENGE PROBLEM

A two-stage secondary treatment facility treats 5.4 MGD or 20.4 ML/d. The first stage of the secondary process is accomplished by trickling filters and the second stage uses activated sludge. The trickling filter effluent is of a secondary treatment quality. The activated sludge process provides for nitrification. The effluent parameters from each stage are given below.

	Trickling Filter Effluent	Activated Sludge Effluent
BOD	28 mg/L	4 mg/L
Suspended Solids	18 mg/L	2 mg/L
Total Phosphorus	0.95 mg/L	0.45 mg/L
Ammonia	95 mg/L	0.32 mg/L

Since the addition of the activated sludge process to the treatment plant operation, the amount of chlorine required for disinfection has more than doubled. What process change could be done to reduce the required amount of chlorine?

CHAPTER 13

Sludge Thickening

QUESTIONS

1. Name four commonly used methods to thicken sludge.

2. Is a gravity thickener better at thickening primary or secondary sludge?

3. Is the efficiency of a gravity thickener improved when the influent to the unit is more concentrated or when it is less concentrated?

4. How can primary sludge be made more fresh going into a gravity thickener?

5. If the biological activity in settling secondary sludge becomes a problem, what two chemicals could be added to help?

6. Why should gravity thickeners be skimmed regularly?

7. Why are pickets sometimes used in a gravity thickener?

8. In a dissolved air flotation unit, what range of pressures is used?

9. What range of sludge concentration can the operator expect to get using a dissolved air flotation thickener?

10. Is it better for the influent to a dissolved air flotation thickener to be more or less concentrated?

11. What range of thickness can the operator expect the sludge float blanket to have when operating a dissolved air flotation thickener?

12. What are the two most important factors that affect the operation of a centrifuge?

13. Name two different spinning parts to a solid bowl centrifuge.

14. What is the differential speed for a solid bowl centrifuge?

15. Is polymer use required with the operation of a gravity belt thickener?

16. What can happen if the size of the holes in the belt of a gravity belt thickener are too small or too large?

PROBLEMS

1. A solid bowl centrifuge receives 48,600 gal of sludge daily. The sludge concentration before thickening is 0.9%. How many pounds of solids are received each day?

2. A solid bowl centrifuge receives 184,000 L of sludge daily. The sludge concentration before thickening is 0.9%. How many kilograms of solids are received each day?

3. A gravity thickener receives a primary sludge flowrate of 162 gpm. If the thickener has a diameter of 26 ft, what is the hydraulic loading rate in gpd/ft²?

4. A gravity thickener receives a primary sludge flowrate of 10.2 L/s. If the thickener has a diameter of 7.9 m, what is the hydraulic loading rate in m³/d/m²?

5. The primary sludge flowrate to a 42 ft diameter gravity thickener is 260 gpm. If the solids concentration is 1.2%, what is the solids loading rate in lbs/d/ft²?

6. The primary sludge flowrate to a 12.8 m diameter gravity thickener is 16.4 L/s. If the solids concentration is 1.2%, what is the solids loading rate in kg/d/m²?

7. Waste activated sludge is pumped to a 32 ft diameter dissolved air flotation thickener at a rate of 680 gpm. What is the hydraulic loading rate in gpm/ft²?

8. Waste activated sludge is pumped to a 9.8 m diameter dissolved air flotation thickener at a rate of 42.5 L/s. What is the hydraulic loading rate in L/s/m²?

9. Waste activated sludge is pumped to a 32 ft diameter dissolved air flotation thickener at a rate of 125 gpm. If the concentration of solids 0.95%, what is the solids loading rate in lbs/hr/ft²?

10. Waste activated sludge is pumped to a 9.8 m diameter dissolved air flotation thickener at a rate of 7.9 L/s. If the concentration of solids is 0.95%, what is the solids loading rate in kg/hr/m²?

11. Sludge is pumped to a gravity thickener at a rate of 147 gpm. Thickened sludge is removed from the thickener at a rate of 77 gpm. The solids in the primary sludge average 2.6% and the solids in the thickened sludge are 5.2%. The gravity thickener overflow has a suspended solids concentration of 384 mg/L.

   a. Do you expect the sludge blanket level to increase, decrease, or remain the same?

   b. If the thickened sludge starts to thin, what can be done?

12. Sludge is pumped to a gravity thickener at a rate of 9.2 L/s. Thickened sludge is removed from the thickener at a rate of 4.8 L/s. The solids in the primary sludge average 2.6% and the solids in the thickened sludge are 5.2%. The gravity thickener overflow has a suspended solids concentration of 384 mg/L.
    a. Do you expect the sludge blanket level to increase, decease, or remain the same?

    b. If the thickened sludge starts to thin, what can be done?

CHAPTER 14

Anaerobic Sludge Digestion

QUESTIONS

1. How many basic steps are there in the anaerobic digestion process?

2. What is the first step in the anaerobic digestion process?

3. What is the second step in the anaerobic digestion process?

4. What are the bugs called that do the first step in the anaerobic digestion process?

5. What are the bugs called that do the second step in the anaerobic digestion process?

6. Describe what pH is.

7. What value of pH is neutral?

8. If the anaerobic digester pH goes below 6.6, what can happen?

9. If the anaerobic digester pH goes above 7.5, what can happen?

10. What is alkalinity?

11. What is the temperature range for mesophilic bugs?

12. What is the temperature range for thermophilic bugs?

13. At what temperature are most anaerobic digesters operated?

14. Digester gas contains what two gases?

15. How is the organic material in the raw sludge measured?

16. What is the explosive range of methane in air?

17. What is the first indication that an anaerobic digester is having a problem?

18. What is the value of the volatile acid to alkalinity ratio for a properly operated anaerobic digester?

19. What is the amount of carbon dioxide in the digester gas for a properly operated anaerobic digester?

20. Name four potential causes of an anaerobic digester operational problem.

21. What is the maximum percent volume that accumulated grit and scum should occupy in a digester?

22. What is the recommended schedule for feeding raw sludge to a digester?

23. What two chemicals can be added to increase alkalinity?

24. If the volatile acid to alkalinity ratio is at a value of 0.23 and rising, is there a problem?

PROBLEMS

1. Sludge is being pumped to the digester at a rate of 3.4 gpm. How many pounds of volatile solids are being pumped to the digester daily if the sludge has a 4.9% total solids content with 69% volatile solids?

2. Sludge is being pumped to the digester at a rate of 18,500 L daily. How many kilograms of volatile solids are being pumped to the digester daily if the sludge has a 4.9% total solids content with 69% volatile solids?

3. A 52 ft diameter anaerobic digester has a liquid depth of 20 ft. The unit receives 46,900 gallons of sludge daily with a solids content of 5.4% of which 70% are volatile. What is the organic loading rate in the digester in lbs VS added/ft³/day?

4. A 15.8 m diameter anaerobic digester has a liquid depth of 6.1 m. The unit receives 139,600 L of sludge daily with a solids content of 5.4% of which 70% are volatile. What is the organic loading rate in the digester in kg VS added/m³/day?

5. The concentration of volatile acids in the anaerobic digester is 176 mg/L. If the concentration of alkalinity is measured to be 2060 mg/L, what is the VA/Alkalinity ratio?

6. If the anaerobic digester becomes sour, it must be neutralized. This can be done by adding lime to the unit. The amount of lime to add is determined by the ratio of one mg/L of lime for every mg/L of volatile acids in the digester. If the volume of sludge in the digester is 196,000 gallons and the volatile acids concentration is 1840 mg/L, how many pounds of lime will be required to neutralize the digester?

7. If the anaerobic digester becomes sour, it must be neutralized. This can be done by adding lime to the unit. The amount of lime to add is determined by the ratio of one mg/L of lime for every mg/L of volatile acids in the digester. If the volume of sludge in the digester is 745,000 liters and the volatile acids concentration is 1840 mg/L, how many kilograms of lime will be required to neutralize the digester?

8. The anaerobic digester has a raw sludge volatile solids content of 70%. The digested sludge has a volatile solids content of 53%. What is the percent reduction in the volatile solids content through the anaerobic digester?

9. The anaerobic digester has a raw sludge volatile solids content of 68 %. The digested sludge has a volatile solids content of 54 %. What is the percent reduction in the volatile solids content through the anaerobic digester?

10. Calculations indicate that 5500 lb of volatile solids will be required in the seed sludge. How many gallons of seed sludge will be required if the sludge has a 9.7 % solids content with 68 % VS and weighs 8.7 lb/gal?

11. Calculations indicate that 2500 kg of volatile solids will be required in the seed sludge. How many liters of seed sludge will be required if the sludge has a 9.7% solids content with 68% VS and weighs 1.04 kg/L?

12. There are 84,000 gal of raw sludge pumped to the digester daily with a solids content of 5.4% and 68% VS. If the VS content of the digested sludge is 52% and the average daily gas production is 12 ft³lb VS destroyed, what is the daily gas production?

13. There are 318,000 L of raw sludge pumped to the digester daily with a solids content of 5.4% and 68% VS. If the VS content of the digested sludge is 52% and the average daily gas production is 0.75 m³/kg VS destroyed, what is the daily gas production?

CHALLENGE PROBLEM

14. Samples from the anaerobic digester are tested for concentrations of volatile acids and alkalinity on Monday, Wednesday, and Friday of each week.

Week	Day	Volatile Acids (mg/L)	Alkalinity (mg/L)
#1	Mon	142	1990
	Wed	145	1830
	Fri	154	1690
#2	Mon	150	1810
	Wed	151	1800
	Fri	149	1910
#3	Mon	150	1850
	Wed	155	1560
	Fri	160	1430
#4	Mon	167	1246
	Wed	174	1140
	Fri	189	1070

a. For each pair of values calculate the VA/Alkalinity ratio.

b. Should the operator of this unit be concerned? What else should be checked? What might be changed?

CHAPTER 15

Sludge Dewatering

QUESTIONS

1. Name three general ways to condition sludge.

2. Name two general ways that conditioned sludge can be dewatered.

3. What are the two basic steps that happen during chemical sludge conditioning?

4. What electrical charge do sludge particles usually have?

5. What happens to the amount of sludge conditioning chemical as the size of the sludge particles gets smaller?

6. Name two popular sludge conditioning chemicals.

7. What happens when lime is mixed with water?

8. Name three advantages of using lime for sludge conditioning.

9. Name two disadvantages of using lime for sludge conditioning.

10. How should a polymer spill be cleaned up?

11. What is the dry polymer's "percent activity" a measure of?

12. What type of charge do anionic polymers have?

13. Why are cationic polymers used most frequently in sludge conditioning?

14. What range of polymer solution concentration is typically used?

15. What happens as a polymer solution is aged?

16. What chemical, when added upstream of the polymer, can reduce the amount of polymer used?

17. What piece of kitchen equipment is most like the thermal conditioning process?

18. What two actions take place on a sludge drying bed?

19. In what type of climates would sludge drying beds work best?

20. How many different belts are used in a belt filter press?

21. Can you name four different areas in a belt filter press? What are they?

22. What chemicals might be added to the sludge to allow the dried solids to let go of the belts more easily?

23. In a solid bowl centrifuge, what two things are spinning?

24. At what differential speed will the sludge cake be the dryest?

25. How much pressure is applied to the sludge in a belt filter press?

26. In order to be effective, a vacuum filter usually requires what type of sludge conditioning?

PROBLEMS

1. Sludge is applied to a drying bed 200 ft long and 22 ft wide. The sludge has a total solids concentration of 3.1% and fills the bed to a depth of 11 in. If it takes an average of twenty-two days for the sludge to dry and one day to remove the dried solids, how many pounds of solids can be dried for every square foot of drying bed area each year?

2. Sludge is applied to a drying bed 61 m long and 6.7 m wide. The sludge has a total solids concentration of 3.1% and fills the bed to a depth of 28 cm. If it takes an average twenty-two days for the sludge to dry and one day to remove the dried solids, how many kilograms of solids can be dried for every square meter of drying bed area each year?

3. A belt filter press receives a daily sludge flow of 0.18 million gal. If the belt is 68 in. wide, what is the hydraulic loading rate on the unit in gallons per minute for each foot of belt width (gpm/ft)?

4. A belt filter press receives a daily sludge flow of 0.68 million L. If the belt is 173 cm wide, what is the hydraulic loading rate on the unit in liters per second for each meter of belt width (L/s/m)?

5. A plate and frame filter press can process 950 gal of sludge during its 145 min operating cycle. If the sludge concentration is 4.0%, and if the plate surface area is 148 ft^2, how many pounds of solids are pressed per hour for each square foot of plate surface area?

6. A plate and frame filter press can process 3600 L of sludge during its 145 min operating cycle. If the sludge concentration is 4.0%, and if the plate surface area is 13.8 m², how many kilograms of solids are pressed per hour for each square meter of plate surface area?

7. Thickened thermally conditioned sludge is pumped to a vacuum filter at a rate of 35 gpm. The vacuum area of the filter is 12 ft wide with a drum diameter of 9.8 ft. If the sludge concentration is 13%, what is the filter yield in lbs/hr/ft²? Assume the sludge weight 8.34 lb/gal.

8. Thickened thermally conditioned sludge is pumped to a vacuum filter at a rate of 2.21 L/s. The vacuum area of the filter is 3.65 m wide with a drum diameter of 3.00 m. If the sludge concentration is 13%, what is the filter yield in kg/hr/m²? Assume the sludge weighs 1 kg/L.

9. The vacuum filter produces an average of 3070 pounds of sludge cake each hour. The total solids content of the cake produced is 45%. The sludge is being pumped to the filter at a rate of 23 gpm and at a concentration of 12%. If the sludge density is 8.51 lb/gal, what is the percent recovery of the filter?

10. The vacuum filter produces an average of 1392 kilograms of sludge cake each hour. The total solids content of the cake produced is 45%. The sludge is being pumped to the filter at a rate of 1.45 L/s and at a concentration of 12%. If the sludge density is 1.02 kg/L, what is the percent recovery of the filter?

CHAPTER 16

Odor Control

QUESTIONS

1. What are the two basic categories of odors in a wastewater treatment plant?

2. What two chemicals are responsible for the majority of the inorganic odors?

3. Which of these chemicals smells like rotten eggs?

4. When hydrogen sulfide is dissolved in water, what kind of ion can it break up into?

5. Is this ion odorous?

6. Above what pH will the hydrogen sulfide stay in this ion type form?

7. What two chemicals can be added to the wastewater to raise the pH?

8. What are the two different forms of ammonia?

9. At a pH below 9.0, what happens to the ammonia in water?

10. Inside what range of pH will the odors of both hydrogen sulfide and ammonia be minimized?

11. How many different organic odors are there?

12. What is the lowest concentration of an odor that the human nose can smell?

13. How can odorous air be sampled?

14. If an odorous air sample is diluted with odor-free air to the point that at least half the people on an odor panel can't smell it, what is this concentration called?

15. What does GC stand for?

16. What does the GC test tell us about an odorous air sample?

17. What does MS stand for?

18. What does the MS test tell us that the GC test can't?

19. What can measure lower concentrations, MS or your nose?

20. Where is the best place to treat odors?

21. What two chemicals are more commonly used in gravity sewers to treat odors associated with hydrogen sulfide?

22. What chemical is recommended for use in force mains?

23. What other chemical has been used to treat hydrogen sulfide?

24. What odors does hydrogen peroxide not treat?

25. Hydrogen sulfide will not form if the DO is kept above what value?

26. What chemical could be added to the wastewater to increase the available oxygen?

27. What kind of odors can scrubbers be effectively used for?

28. What kind of odors will scrubbers not treat very well?

29. What two methods are used to effectively treat organic odors?

30. What is a masking agent?

PROBLEMS

1. It has been decided to treat the headworks odor control problems by dosing sodium nitrate ($NaNO_3$) at a rate of 10 lbs per pound of influent sulfide. The influent sulfide concentration average is 0.2 mg/L. The plant treats an average daily flowrate of 4.5 MGD.

   a. How many 40-lb bags of Na NO_3 would the facility use in a week?

b. What alternatives might the plant consider?

2. It has been decided to treat the headworks odor control problems by dosing sodium nitrate ($NaNO_3$) at a rate of 10 kg per kilogram of influent sulfide. The influent sulfide concentration averages 0.2 mg/L. The plant treats an average daily flowrate of 17.0 ML/d.

a. How many 18-kg bags of $NaNO_3$ would the facility use in a week?

b. What alternatives might the plant consider?

3. Calcium hypochlorite is added to the city's sewer collection system to reduce potential odors. At a particular location the chemical is dosed at a rate of 20 mg/L. The flowrate in this particular interceptor is 0.6 MGD. The calcium hypochlorite has 60% available chlorine by weight.
    a. How many pounds of calcium hypochlorite would be required for a 30-day supply?

b. What change might you recommend?

4. Calcium hypochlorite is added to the city's sewer collection system to reduce potential odors. At a particular location the chemical is dosed at a rate of 20 mg/L. The flowrate in this particular interceptor is 2270 m^3/d. The calcium hypochlorite has 60% available chlorine by weight.
   a. How many kilograms of calcium hypochlorite would be required for a 30-day supply?

b. What change might you recommend?

5. A study has been conducted concerning an odor control problem with the treatment plant headworks. The results of the study found no specific industrial discharge contributing to this problem. Also, the gravity sewer system has no pumping stations and no feasible place to dose chemical prior to the headworks. The ammonia and hydrogen sulfide odors have resulted in many complaints from the community. The plant influent has 12 mg/L of hydrogen sulfide. It has been decided to dose hydrogen peroxide at a rate of 2.5 parts H_2O_2 per part H_2S. The plant treats an average daily flowrate of 1.0 MGD.

a. How many gallons of 75% H_2O_2 solution will be needed per day?

b. How effective should this treatment be?

6. A study has been conducted concerning an odor control problem with the treatment plant headworks. The results of the study found no specific industrial discharge contributing to this problem. Also, the gravity sewer system has no pumping stations and no feasible place to dose chemical prior to the headworks. The ammonia and hydrogen sulfide odors have resulted in many complaints from the community. The plant influent has 12 mg/L of hydrogen sulfide. It has been decided to dose hydrogen peroxide at a rate of 2.5 parts H_2O_2 per part H_2S. The plant treats an average daily flowrate of 3785 m^3/d.
   a. How many liters of 75% of H_2O_2 solution will be needed per day?

b. What else can be done to reduce the odor?

CHAPTER 17

Lab

QUESTIONS

1. Describe what pH is.

2. Give one way to describe an acid.

3. What is an example of a chemical that is a base?

4. If an acid solution is mixed with a base solution so that they neutralize each other, what is left in the solution?

5. A neutral solution has what value of pH?

6. What is the concentration of hydrogen ions at a pH of 7.0?

7. What is the concentration of hydrogen ions at a pH of 6.0?

8. What is the concentration of hydrogen ions at a pH of 5.0?

9. What happens to the concentration of hydrogen ions as the pH goes down by 1.0 units?

10. What chemical is used to supply oxygen in the COD test?

11. How large a value is the BOD as compared to the COD?

12. What is the biggest advantage of the COD over the BOD?

13. What is the most common form of phosphorus in wastewaters?

14. What happens chemically during the first step of a test for total phosphorus concentration?

15. After the samples being tested for phosphorus turn blue, what machine is used to find the exact phosphorus concentrations?

16. What bug is used most frequently as an indicator organism?

17. What chemical is often used to dechlorinate an effluent sample before doing biological testing?

18. What test is most commonly used to count the indicator bugs in the plant effluent?

19. What would you call a dense population of bugs grown on one spot where only one bug has started?

20. In a membrane filter test for fecal coliform there are black, blue, and yellow dots. Which ones should be counted?

21. When a membrane filter test for fecal coliform is run, there is more than one sample volume used. Which results are used for the calculation?

22. How are the results of a membrane filter test reported?

23. Why is a seeded BOD test needed for some samples?

24. For a chlorinated effluent sample, what is the best source of "seed" bugs to use?

25. What is the name given to the final DO in a BOD test?

26. What is the name given to the difference between the initial and final DOs in a BOD test?

27. What is the name given to the amount of oxygen used during a BOD test?

28. Describe what the depletion for the seed-alone sample represents.

29. Describe what the depletion for the seeded sample represents.

30. If you found the difference between the depletions described in questions 28 and 29, what would this represent?

31. Is the amount of oxygen used by the bugs in the seeded sample to eat only seed food measured directly?

32. What is a "blank" in the BOD test?

33. Is ammonia added to a BOD test bottle?

34. If nitrifiers are in a sample, will they eat the ammonia added to a BOD bottle?

35. If nitrifiers are in a sample, is the BOD test accurate?

36. If a chemical is added to the BOD bottle to prevent the nitrifiers from growing, what is the test called?

37. What is NOD?

38. Name two ways lab chemical concentrations may be given.

39. What is molarity?

40. What is a burette?

41. What does GAW stand for?

42. What is the equivalent weight of an acid?

43. What is the equivalent weight of a base?

44. What is normality?

PROBLEMS

1. Three different volumes of an effluent sample were tested for fecal coliform.
   The counts for each were:

Volume:	25 mL	10 mL	5 mL
Count:	108	51	25

   Calculate the fecal coliform count.

2. Three different volumes of an effluent sample were tested for fecal coliform. The incubated filters are shown below:

Count the filters and calculate the fecal coliform count per 100 mL.

3. A BOD test is done on a primary effluent sample. The initial DO was 6.94 mg/L and the residual DO was 4.36 mg/L. If the sample percentage used was 5%, what was the BOD of the sample?

4. A BOD test is done on a primary effluent sample. The initial DO was 7.43 mg/L and the residual DO was 4.21 mg/L. If the sample volume used was 10 mL, what was the BOD of the sample? Note: A BOD bottle holds 300 mL.

5. Use the following information to calculate the seeded BOD concentration.

Seeded Effluent:

$DO_i = 7.29$ mg/L
$DO_f = 4.62$ mg/L
%Eff $= 18\%$
%Seed $= 0.6\%$

Seed Alone:

$DO_i = 7.58$ mg/L
$DO_f = 4.89$ mg/L
%Seed $= 4\%$

6. Use the following information to calculate the seeded BOD concentration.

      Seeded Effluent: Seed Alone:

      $DO_i = 6.86$ mg/L $DO_i = 7.44$ mg/L
      $DO_f = 3.95$ mg/L $DO_f = 4.62$ mg/L
      %Eff = 25%   %Seed = 5%
      %Seed = 0.5%

On some operator certification exams the initial DOs must be calculated from the DOs of the dilution water, sample, and seed. For this reason, the following problems are included.

7. Calculate the initial DOs for the seeded sample and the seed alone sample from the following information.

    Dilution Water DO = 8.1 mg/L
    Effluent Sample DO = 5.2 mg/L
    Seed DO = 2.0 mg/L
    % Effluent = 12%
    % Seed with Effluent = 1%
    % Seed Alone = 6%

8. Calculate the initial DOs for the seeded sample and the seed alone sample from the following information.

Dilution Water DO = 7.9 mg/L
Effluent Sample DO = 4.8 mg/L
Seed DO = 0.0 mg/L
% Effluent = 18%
% Seed with Effluent = 0.6%
% Seed Alone = 4%

9. Calculate the seeded BOD concentration from the following information.

$$
\begin{aligned}
\text{Dilution Water DO} &= 8.1 \text{ mg/L} \\
\text{Effluent Sample DO} &= 4.6 \text{ mg/L} \\
\text{Seed DO} &= 1.5 \text{ mg/L} \\
\text{Effluent Residual DO} &= 4.23 \text{ mg/L} \\
\text{Seed Alone Residual DO} &= 4.92 \text{ mg/L} \\
\text{Blank Residual DO} &= 8.1 \text{ mg/L} \\
\% \text{ Effluent} &= 18\% \\
\% \text{ Seed with Effluent} &= 1\% \\
\% \text{ Seed Alone} &= 8\%
\end{aligned}
$$

10. Calculate the seeded BOD concentration from the following information.

$$
\begin{aligned}
\text{Dilution Water DO} &= 7.96 \text{ mg/L} \\
\text{Effluent Sample DO} &= 2.43 \text{ mg/L} \\
\text{Seed DO} &= 0.00 \text{ mg/L} \\
\text{Effluent Residual DO} &= 3.98 \text{ mg/L} \\
\text{Seed Alone Residual DO} &= 4.12 \text{ mg/L} \\
\text{Blank Residual DO} &= 7.96 \text{ mg/L} \\
\% \text{ Effluent} &= 12\% \\
\% \text{ Seed with Effluent} &= 0.4\% \\
\% \text{ Seed Alone} &= 5\%
\end{aligned}
$$

11. A 2-L volume of 0.05 N hydrochloric acid (HCl) solution is to be prepared. How many milliliters of 10 N hydrochloric acid must be diluted with water to prepare the desired solution?

12. It takes 8.4 mL of a solution of HCl to neutralize 10 mL of 5N NaOH. What is the concentration of the HCl solution?

CONVERSION FACTORS

Quantity	Equivalent Values
Mass	1 kg = 1000 g = 0.001 metric ton = 2.205 lb = 35.274 oz 1 lb = 16 oz = 453.6 g = 0.4536 kg
Length	1 m = 100 cm = 1000 mm = 39.37 in. = 3.281 ft = 1.094 yd 1 ft = 12 in. = 1/3 yd = 0.3048 m = 30.48 cm
Volume	$1 m^3$ = 1000 L = 106 cm^3 = 10^6 ml = 35.3145 ft^3 = 264.17 gal = 1056.68 qt 1 ft^3 = 1728 $in.^3$ = 7.4805 gal = 0.02832 m^3 = 28.317 L

Some Common Wastewater Conversion Factors
1 ft^3 = 7.48 gal
1 gal = 8.34 lb (water) 1 Liter = 1 kg
1 MGD = 3785.4 m^3/d
= 3.7854 ML/d
1 L/s = 15.851 gpm
1 ac = 43,560 ft^2
1 ha = 10,000 m^2 = 2.471 ac

REFERENCES

The primary resources supporting the technical information presented in this book include the following:

1 *Activated Sludge Microbiology*, Richard (1989).
2 *Applied Math for Wastewater Plant Operators*, Price (1991).
3 *Biological Wastewater Treatment*, Grady & Lim (1980).
4 *MOP 11, Second Edition*, Water Environment Federation (1990).
5 *Operation of Wastewater Treatment Plants, 3rd Edition*, California State U. (1986).
6 *Standard Methods for the Analysis of Water and Wastewater, 17th Edition*, Am. Public Health Assoc. (1989).
7 *Wastewater Engineering*, Metcalf & Eddy (1979).
8 *Wastewater Treatment Plants, Planning, Design & Operation*, Qasim (1985).

CHAPTER 18

Answers to the Questions – Solutions to the Problems

ANSWERS TO THE QUESTIONS: INTERCEPTORS AND WET WELL PUMPING

1. The wastewater that comes from the household is called domestic waste.

2. The wastewater that comes from manufacturing plants is called industrial waste.

3. If the domestic waste and industrial waste are combined the wastewater is called municipal waste.

4. A sewer that contains only municipal waste is called a sanitary sewer.

5. A sewer that contains only storm water is called a storm sewer.

6. A sewer that contains both municipal waste and storm water is called a combined sewer.

7. If an interceptor with both municipal waste and storm water overflows into a receiving stream, the event is called a combined sewer overflow or CSO.

8. One problem caused by loose sewer line joints is ground water infiltration.

9. An increase in the amount of sand coming into the plant might be an indication of a broken sewer line.

10. The ideal velocity for flow to go through an interceptor is 2 ft/s or 0.6 m/s.

SOLUTIONS TO THE PROBLEMS: INTERCEPTORS AND WET WELL PUMPING

1.
$$\text{Influent} = \text{Discharge} + \text{Accumulation}$$

$$0 \text{ gpm} = \text{Discharge} + \text{Accumulation}$$

$$\text{Accumulated Volume} = 10 \text{ ft} \times 12 \text{ ft} \times -2.1 \text{ ft} \times \frac{7.48 \text{ gal}}{1 \text{ ft}^3}$$

$$= -1885 \text{ gal}$$

$$\text{Accumulation} = \frac{1885 \text{ gal}}{5 \text{ min}} = \frac{377 \text{ gal}}{1 \text{ min}} = -377 \text{ gpm}$$

$$0 \text{ gpm} = \text{Discharge} + (-377 \text{ gpm})$$

$$377 \text{ gpm} = \text{Discharge}$$

2.
$$\text{Influent} = \text{Discharge} + \text{Accumulation}$$

$$0 \text{ L/s} = \text{Discharge} + \text{Accumulation}$$

$$\text{Accumulated Volume} = 3.1 \text{ m} \times 3.6 \text{ m} \times -0.7 \text{ m} \times \frac{1000 \text{ L}}{1 \text{ m}^3}$$

$$= -7812 \text{ L}$$

$$\text{Time} = 5 \text{ minutes} \times \frac{60 \text{ s}}{1 \text{ min}} = 300 \text{ s}$$

$$\text{Accumulation} = \frac{-7812 \text{ L}}{300 \text{ s}} = \frac{-26 \text{ L}}{1 \text{ s}} = -26 \text{ L/s}$$

$$0 \text{ L/s} = \text{Discharge} + (-26 \text{ L/s})$$

$$26 \text{ L/s} = \text{Discharge}$$

3.
$$\text{Accumulation} = \frac{11 \text{ ft} \times 13 \text{ ft} \times -1.8 \text{ ft}}{4 \text{ min}} \times \frac{7.48 \text{ gal}}{1 \text{ ft}^3}$$

$$= -481 \text{ gpm}$$

$$\text{Influent} = \text{Discharge} + \text{Accumulation}$$

$$0 \text{ gpm} = \text{Discharge} + (-481 \text{ gpm})$$

$$481 \text{ gpm} = \text{Discharge}$$

4.

$$\text{Time} = 4 \text{ min} \times \frac{60 \text{ s}}{\text{min}} = 240 \text{ s}$$

$$\text{Accumulation} = \frac{3.4 \text{ m} \times 4 \text{ m} \times -0.5 \text{ m}}{240 \text{ seconds}} \times \frac{1000 \text{ L}}{1 \text{ m}^3}$$

$$= -28.3 \text{ L/s}$$

$$\text{Influent} = \text{Discharge} + \text{Accumulation}$$

$$0 \text{ L/s} = \text{Discharge} + (-28.3 \text{ L/s})$$

$$28.3 \text{ L/s} = \text{Discharge}$$

5. Before we start, all the units need to be put into feet and minutes.

$$10 \text{ ft-9 in.} = 10 \text{ ft} + 9 \text{ in.} \times \frac{1 \text{ ft}}{12 \text{ in}} = 10.75 \text{ ft}$$

$$12 \text{ ft-2 in.} = 12 \text{ ft} + 2 \text{ in.} \times \frac{1 \text{ ft}}{12 \text{ in}} = 12.17 \text{ ft}$$

$$6 \text{ min-35 s} = 6 \text{ min} + 35 \text{ s} \times \frac{1 \text{ min}}{60 \text{ sec}} = 6.58 \text{ min}$$

$$\text{Accumulation} = \frac{10.75 \text{ ft} \times 12.17 \text{ ft} \times -3 \text{ ft}}{6.58 \text{ min}} \times \frac{7.48 \text{ gal}}{1 \text{ ft}^3}$$

$$= -446 \text{ gpm}$$

$$\text{Influent} = \text{Discharge} + \text{Accumulation}$$

$$0 \text{ gpm} = \text{Discharge} + (-446 \text{ gpm})$$

$$446 \text{ gpm} = \text{Discharge}$$

6. Before we start, all the units need to be put into meters and seconds.

$$328 \text{ cm } \times \frac{1 \text{ m}}{100 \text{ cm}} = 3.28$$

$$372 \text{ cm } \times \frac{1 \text{ m}}{100 \text{ cm}} = 3.72 \text{ m}$$

$$50 \text{ cm } \times \frac{1 \text{m}}{100 \text{ cm}} = 0.50 \text{ m}$$

$$3 \text{ min-}37 \text{ s} = \frac{(3 \text{ min} \times 60 \text{ s})}{1 \text{ min}} + 37 \text{ s} = 217 \text{ s}$$

$$\text{Accumulation} = \frac{3.28 \text{ m} \times 3.72 \text{ m} \times -0.50 \text{ m}}{217 \text{ s}} \times \frac{1000 \text{ L}}{1 \text{ m}^3}$$

$$= -28.1 \text{ L/s}$$

Influent = Discharge + Accumulation

0 L/s = Discharge + (−28.1 L/s)

28.1 L/s = Discharge

---

7.   Influent = Discharge + Accumulation

370 gpm = Discharge + Accumulation

$$\text{Accumulation} = \frac{11.5 \text{ ft} \times 13 \text{ ft} \times 1 \text{ in.}}{7 \text{ min}} \times \frac{1 \text{ ft}}{12 \text{ in}} \times \frac{7.48 \text{ gal}}{1 \text{ ft}^3}$$

$$= 13.3 \text{ gpm}$$

Influent = Discharge + Accumulation

370 gpm = Discharge + 13.3 gpm

356.7 gpm = Discharge

---

8.   Influent = Discharge + Accumulation

24 L/s = Discharge + Accumulation

$$\text{Accumulation} = \frac{3.3 \text{ m} \times 4.1 \text{ m} \times 3 \text{ cm}}{6 \text{ min} \times 60 \text{ s/min}} \times \frac{1 \text{ m}}{100 \text{ cm}} \times \frac{1000 \text{ L}}{1 \text{ m}^3}$$

$$= 1.1 \text{ L/s}$$

$$\text{Influent} = \text{Discharge} + \text{Accumulation}$$

$$24 \text{ L/s} = \text{Discharge} + 1.1 \text{ L/s}$$

$$22.9 \text{ L/s} = \text{Discharge}$$

9.

$$130 \text{ in.} = \frac{1 \text{ ft}}{12 \text{ in.}} = 10.83$$

$$146 \text{ in.} = \frac{1 \text{ ft}}{12 \text{ in.}} = 12.17 \text{ ft}$$

$$\text{Accumulation} = \frac{10.83 \text{ ft} \times 12.17 \text{ ft} \times (-1.5 \text{ in.})}{6 \text{ min}} \times \frac{1 \text{ ft}}{12 \text{ in.}}$$

$$\times \frac{7.48 \text{ gal}}{1 \text{ ft}^3} = (-20.5 \text{ gpm})$$

$$\text{Influent} = \text{Discharge} + \text{Accumulation}$$

$$450 \text{ gpm} = \text{Discharge} + (-20.5 \text{ gpm})$$

$$470.5 \text{ gpm} = \text{Discharge}$$

10.

$$335 \text{ cm} \times \frac{1 \text{ m}}{100 \text{ cm}} = 3.35 \text{ m}$$

$$370 \text{ cm} \times \frac{1 \text{ m}}{100 \text{ cm}} = 3.70 \text{ m}$$

$$\text{Accumulation} = \frac{3.35 \text{ m} \times 3.70 \text{ m} \times (-2.5 \text{ cm})}{7 \text{ min} \times 60 \text{ sec/min}} \times \frac{1 \text{ m}}{100 \text{ cm}}$$

$$\times \frac{1000 \text{ L}}{1 \text{ m}^3} = (-0.7 \text{ L/s})$$

$$\text{Influent} = \text{Discharge} + \text{Accumulation}$$

$$30 \text{ L/s} = \text{Discharge} + (-0.7 \text{ L/s})$$

$$30.7 \text{ L/s} = \text{Discharge}$$

---

11. $$\text{Total Accumulation} = \frac{9.5 \text{ ft} \times 13 \text{ ft} \times (-7 \text{ in.})}{14 \text{ min}} \times \frac{1 \text{ ft}}{12 \text{ in.}}$$

$$\times \frac{7.48 \text{ gal}}{1 \text{ ft}^3} = (-38.5 \text{ gpm})$$

Influent = Discharge + Accumulation

750 gpm = Discharge + (−38.5 gpm)

788.5 gpm = Discharge

Discharge = First Pump + Second Pump

788.5 gpm = 490 gpm + Second Pump

298.5 gpm = Second Pump Discharge

---

12. $$\text{Total Accumulation} = \frac{2.9 \text{ m} \times 4 \text{ m} \times (-12 \text{ cm})}{15 \text{ min} \times 60 \text{ s/min}} \times \frac{1 \text{ m}}{100 \text{ cm}}$$

$$\times \frac{1000 \text{ L}}{1 \text{ m}^3} = (-1.5 \text{ L/s})$$

Influent = Discharge + Accumulation

47.2 L/s = Discharge + (−1.5 L/s)

48.7 L/s = Discharge

Discharge = First Pump + Second Pump

48.7 L/s = 30 L/s + Second Pump

18.7 L/s = Second Pump Discharge

ANSWERS TO THE QUESTIONS: PRELIMINARY TREATMENT

1. The screenings can include rags, cans, leaves, and branches.

2. A plant usually removes between 0.5 and 12 ft³ of screenings per million gallons of wastewater received.

3. A plant usually removes between 3.5 and 80 m³ of screenings per million cubic meters of wastewater received.

4. The two types of coarse influent screens are bar racks and bar screens.

5. A bar screen has bar spacings ranging from 0.5 to 2 in. A bar rack can have bar spacings anywhere from 2 to 4 in.

6. A mechanical rake can be operated manually, based on a timer, or automatically based on channel level.

7. Two devices used to monitor water level are the bubbler and the sonic level sensor.

8. The optimum velocity through a bar screen is 1.5 ft/s or 0.5 m/s.

9. Grit will settle in the screening channel if the velocity gets down to 1 ft/s or 0.3 m/s.

10. Approximately 80% of the screenings are water.

11. Grit can contain sand and gravel as well as coffee grounds and egg shells.

12. Reasons we need to remove the grit early in the treatment process is because of the excessive wear that grit causes to pumps and because it will settle and take up valuable space in downstream units.

13. The two basic kinds of grit removal units are the grit chamber and the aerated grit channel.

14. In order to settle the grit the wastewater velocity should be slowed down to 1 ft/s or 0.3 m/s.

15. Too much air in an aerated grit channel will cause the wastewater to roll too fast. This will then cause the grit to stay suspended in the water and be carried into the next unit.

16. The shape of the water path through an aerated grit channel is a spiral.

SOLUTIONS TO THE PROBLEMS: PRELIMINARY TREATMENT

1. $\dfrac{4.8 \text{ million gal}}{1 \text{ d}} \times \dfrac{1 \text{ d}}{24 \text{ hr}} \times 96 \text{ hr} = 19.2 \text{ million gal} = 19.2 \text{ MG}$

$$\dfrac{4.2 \text{ ft} \times 5.6 \text{ ft} \times 26 \text{ in.}}{19.2 \text{ MG}} \times \dfrac{1 \text{ ft}}{12 \text{ in.}} = \dfrac{2.7 \text{ ft}^3}{1 \text{ MG}}$$

2. The unit ML represents million liters. We need to convert this to cubic meters.

$$\dfrac{20 \text{ ML}}{1 \text{ d}} \times \dfrac{20,000,000 \text{ L}}{1 \text{ d}} \times \dfrac{1 \text{ m}^3}{1000 \text{ L}} = \dfrac{20,000 \text{ m}^3}{1 \text{ d}}$$

$$\dfrac{20,000 \text{ m}^3}{1 \text{ d}} \times 4 \text{ d} = 80,000 \text{ m}^3$$

We need to convert m^3 to million cubic meters.

$$80,000 \text{ m}^3 \times \dfrac{1 \text{ million m}^3}{1,000,000 \text{ m}^3} = \dfrac{80,000}{1,000,000} \text{ million m}^3$$

$$= 0.08 \text{ million m}^3 = 0.08 \text{ million cubic meters}$$

In English units million gallons is represented by MG. In metric units million liters is represented by ML. However, it may be confusing to represent a million cubic meters Mm^3. Instead it is more customary to use scientific notation. Million cubic meters is represented by 10^6m^3. 0.08 million cubic meters = $0.08\ 10^6m^3$.

Returning to the problem:

$$\frac{1.23\ m\ \times\ 1.54\ m\ \times\ 95\ cm}{0.08\ 10^6m^3} \times \frac{1\ m}{100\ cm} = 22.5\ m^3/10^6m^3$$

3. First convert the flowrate to an expression in cubic feet.

$$\frac{650\ gal}{1\ min} \times \frac{1\ ft^3}{7.48\ gal} = \frac{86.9\ ft^3}{1\ min}$$

We want the velocity in feet per second so the time reference must be changed. Finally, the flowrate in cubic feet must be divided by the water width and the water depth to see how fast our imaginary cardboard piece is going through the water.

$$\frac{86.9\ ft^3}{1\ min} \times \frac{1\ min}{60\ sec} \times \frac{1}{(1.1\ ft\ depth\ \times\ 1.3\ ft\ width)} = 1.0\ ft/s$$

4. First convert the flowrate to an expression in cubic meters.

$$\frac{40\ L}{1\ sec} \times \frac{1\ m^3}{1000\ L} = \frac{0.04\ m^3}{1\ sec}$$

Finally, the flowrate in cubic meters must be divided by the water width and the water depth to see how fast our imaginary cardboard piece is going through the water.

$$\frac{0.04\ m^3}{1\ sec} \times \frac{1}{0.32\ m\ depth\ \times\ 0.41\ m\ width} = 0.30\ m/s$$

5. First convert the flowrate in MGD to gpm

$$0.9\ MGD = \frac{900,000\ gal}{1\ d} \times \frac{1\ d}{24\ hr} \times \frac{1\ hr}{60\ min} = \frac{625\ gal}{1\ min} = 625\ gpm$$

Next convert the width and depth in inches to feet.

$$14\ in. \times \frac{1\ ft}{12\ in.} = 1.17\ ft$$

$$16 \text{ in.} \times \frac{1 \text{ ft}}{12 \text{ in.}} = 1.33 \text{ ft}$$

Then convert the flowrate to ft³/s

$$\frac{625 \text{ gal}}{1 \text{ min}} \times \frac{1 \text{ ft}^3}{7.48 \text{ gal}} \times \frac{1 \text{ min}}{60 \text{ sec}} = \frac{1.39 \text{ ft}^3}{1 \text{ sec}}$$

Finally:

$$\frac{1.39 \text{ ft}^3}{1 \text{ s}} \times \frac{1}{1.17 \text{ ft} \times 1.33 \text{ ft}} = \frac{0.89 \text{ ft}}{1 \text{ s}} = 0.89 \text{ ft/s}$$

Note: All this attention to unit cancellation allows the following type of expression:

$$\frac{\text{ft}}{\text{s}} = \frac{900{,}000 \text{ gal}}{1 \text{ d}} \left| \frac{1 \text{ d}}{24 \text{ hr}} \right| \frac{1 \text{ hr}}{60 \text{ min}} \left| \frac{1 \text{ ft}^3}{7.48 \text{ gal}} \right| \frac{1 \text{ min}}{60 \text{ s}} \left| \frac{}{14 \text{ in}} \right|$$

$$\left| \frac{12 \text{ in.}}{1 \text{ ft}} \right| \frac{}{16 \text{ in.}} \left| \frac{12 \text{ in.}}{1 \text{ ft}} = \frac{0.89 \text{ ft}}{1 \text{ s}}\right.$$

The units serve as a map to tell you where the numbers should go.

---

6. First convert the flowrate to cubic meters per second.

$$\frac{39.8 \text{ L}}{1 \text{ s}} \times \frac{1 \text{ m}^3}{1000 \text{ L}} = \frac{0.0398 \text{ m}^3}{1 \text{ s}}$$

Next convert width and depth to units of meters.

$$31.5 \text{ cm} \times \frac{1 \text{ m}}{100 \text{ cm}} = 0.315 \text{ m}$$

$$41.2 \text{ cm} \times \frac{1 \text{ m}}{100 \text{ cm}} = 0.412 \text{ m}$$

Finally:

$$\frac{0.0398 \text{ m}^3}{1 \text{ s}} \times \frac{1}{(0.315 \text{ m} \times 0.412 \text{ m})} = \frac{0.31 \text{ m}}{1 \text{ s}} = 0.31 \text{ m/s}$$

Alternate solution:

$$\frac{m}{s} = \frac{39.8\ L}{1\ s}\ \bigg|\ \frac{1\ m^3}{1000\ L}\ \bigg|\ \frac{}{31.5\ cm}\ \bigg|\ \frac{100\ cm}{1\ m}\ \bigg|\ \frac{}{41.2\ cm}\ \bigg|\ \frac{100\ cm}{1\ m}$$

$$= \frac{0.31\ m}{1\ s}$$

ANSWERS TO THE QUESTIONS: PLANT LOADINGS

1. The ceramic filter holder used in the suspended solids test is called a gooch crucible.

2. Filters are dried at a temperature between 103 and 105°C.

3. Biological activity is measured by the BOD or biochemical oxygen demand test.

4. The BOD bottle contains exactly 300 milliliters (mL).

5. The bugs eat the food in the water. They need oxygen in order to do that. The amount of oxygen used is a measure of how much food the bugs eat.

6. BOD samples are incubated for 5 days.

7. A realistic maximum initial DO for a BOD test is 9 mg/L.

SOLUTIONS TO THE PROBLEMS: PLANT LOADINGS

1. Crucible and solids   25.6829 g
    Crucible (before) -25.6732 g
                       0.0097 g

$$0.0097 \text{ g} \times \frac{1000 \text{ mg}}{1 \text{ g}} = 9.7 \text{ mg}$$

$$50 \text{ mL} \times \frac{1 \text{ L}}{1000 \text{ mL}} = 0.05 \text{ L}$$

$$\frac{9.7 \text{ mg}}{0.05 \text{ L}} = \frac{194 \text{ mg}}{1 \text{ L}} = 194 \text{ mg/L}$$

2. Crucible and solids   26.1425 g
   Crucible (before) $\underline{-26.1349 \text{ g}}$
                        0.0076 g

$$0.0076 \text{ g} \times \frac{1000 \text{ mg}}{1 \text{ g}} = 7.6 \text{ mg}$$

$$25 \text{ mL} \times \frac{1 \text{ L}}{1000 \text{ mL}} = 304 \text{ mg/L}$$

$$\frac{7.6 \text{ mg}}{0.025 \text{ L}} = \frac{304 \text{ mg}}{1 \text{ L}} = 304 \text{ mg/L}$$

3.   Initial DO $(DO_i) = 7.82 \text{ mg/L}$

                 Final DO $(DO_f) = 4.17 \text{ mg/L}$

                 DO Depletion $= 3.65 \text{ mg/L}$

The BOD bottle is 300 mL. Therefore, the sample fraction was:

$$\frac{5 \text{ mL}}{300 \text{ mL}} = 0.0167$$

$$\text{BOD} = \frac{\text{DO Depletion}}{\text{Sample Fraction}} = \frac{3.65 \text{ mg/L}}{0.0167} = 218.6 \text{ mg/L}$$

Note: BODs are always rounded off to whole numbers. BOD = 219 mg/L.

You could actually use the sample volume in milliliters directly in the calculation instead of converting to a sample fraction.

$$\text{BOD} = \text{DO Depletion} \times \frac{300 \text{ mL}}{\text{Sample mL}}$$

$$BOD = 3.65 \text{ mg/L} \times \frac{300 \text{ mL}}{5 \text{ mL}} = 219 \text{ mg/L}$$

4.
$$BOD = DO \text{ Depletion} \times \frac{300 \text{ mL}}{\text{Sample mL}}$$

$$BOD = (7.84 \text{ mg/L} - 4.08 \text{ mg/L}) \times \frac{300 \text{ mL}}{5 \text{ mL}}$$

$$= 226 \text{ mg/L}$$

5.
$$\frac{14.3 \text{ parts}}{\text{million parts}} \times \frac{2.74 \text{ million gal}}{1 \text{ d}} \times \frac{8.34 \text{ lb}}{1 \text{ gal}}$$

$$= \frac{326.8 \text{ lb}}{1 \text{ d}}$$

$$14.3 \text{ mg/L} \times 2.74 \text{ MGD} \times \frac{8.34 \text{ lb}}{1 \text{ gal}} = \frac{326.8 \text{ lb}}{1 \text{ d}}$$

6.
$$\frac{14.3 \text{ mg}}{1 \text{ L}} \times \frac{1 \text{ kg}}{1,000,000 \text{ mg}} \times \frac{9350 \text{ m}^3}{1 \text{ d}} \times \frac{1000 \text{ L}}{1 \text{ m}^3} = \frac{133.7 \text{ kg}}{1 \text{ d}}$$

$$\frac{14.3 \text{ mg/L} \times 9350 \text{ m}^3/\text{d}}{1000} = 133.7 \text{ kg/d}$$

7.
$$290 \text{ mg/L} \times 5.70 \text{ MGD} \times \frac{8.34 \text{ lb}}{1 \text{ gal}} = \frac{13{,}786 \text{ lb}}{1 \text{ d}}$$

8.
$$\frac{290 \text{ mg/L} \times 19{,}500 \text{ m}^3/\text{d}}{1000} = 5655 \text{ kg/d}$$

9.
$$192 \text{ mg/L} \times 4.65 \text{ MGD} \times \frac{8.34 \text{ lb}}{1 \text{ gal}} = \frac{7446 \text{ lb}}{1 \text{ d}}$$

10.
$$\frac{192 \text{ mg/L} \times 17{,}600 \text{ m}^3/\text{d}}{1000} = 3379 \text{ kg/d}$$

ANSWERS TO THE QUESTIONS: PRIMARY TREATMENT

1. The main purpose of a primary clarifier is to remove settleable solids.

2. Settleable solids are measured by pouring a sample into a clear graduated Imhoff Cone. The settleable solids are those below the clear liquid after 60 minutes of settling. They is measured as mL per L sample.

3. Primary settling removes 90 to 95% of the settleable solids.

4. A sample tested for total solids is poured into a ceramic bowl and dried for 24 h. The weight of the dried solids is used to calculate the concentration of total solids.

5. Only 10 to 15% of the total solids are removed by primary settling.

6. Primary settling will typically remove between 25 and 35% of influent BOD.

7. Anywhere from 40 to 60% of the suspended solids are removed by primary settling.

8. A typical detention time in a primary clarifier is 1.5 to 2.5 h.

9. The two chemicals that wastewater treatment plants try to reduce in concentration are phosphorus and ammonia.

10. Secondary treatment follows primary treatment.

SOLUTIONS TO THE PROBLEMS: PRIMARY TREATMENT

1.
$$\text{Tank Volume} = 82 \text{ ft} \times 20 \text{ ft} \times 12.5 \text{ ft} \times \frac{7.48 \text{ gal}}{1 \text{ ft}^3}$$

$$= 153,340 \text{ gal}$$

$$\text{Detention Time} = 153,340 \text{ gal} \times \frac{1 \text{ d}}{1,860,000 \text{ gal}} \times \frac{24 \text{ hr}}{1 \text{ d}} = 2.0$$

2.
$$\text{Tank Volume} = 25 \text{ m} \times 6.1 \text{ m} \times 3.8 \text{ m} = 579.5 \text{ m}^3$$

$$\text{Detention Time} = 579.5 \text{ m}^3 \times \frac{1 \text{ d}}{7041 \text{ m}^3} \times \frac{24 \text{ hr}}{1 \text{ day}} = 2.0 \text{ hr}$$

3.
$$\text{Surface} = 80 \text{ ft} \times 18 \text{ ft} = 1440 \text{ ft}^2$$

$$\text{SOR} = \frac{1,350,000 \text{ gpd}}{1440 \text{ ft}^2} = \frac{938 \text{ gpd}}{1 \text{ ft}^2}$$

4.
$$\text{Surface} = 23 \text{ m} \times 5.4 \text{ m} = 124.2 \text{ m}^2$$

$$\text{SOR} = \frac{5062 \text{ m}^3/\text{d}}{124.2 \text{ m}^2} = \frac{40.8 \text{ m}^3/\text{d}}{1 \text{ m}^2}$$

5. Sample + Dish 74.56 g   Dish + Dry Solids 22.07 g
   Dish Alone −21.34 g   Dish Alone −21.34 g
   Sample Weight 53.22 g   Dry Solids Weight 0.73 g

$$\frac{0.73 \text{ g}}{53.22 \text{ g}} \times 100\% = 1.4\%$$

6.
$$\frac{390 \text{ gal}}{1 \text{ min}} \times \frac{60 \text{ min}}{1 \text{ hr}} \times \frac{24 \text{ hr}}{1 \text{ d}} \times \frac{8.34 \text{ lb}}{1 \text{ gal}} \times \frac{0.9\%}{100.0\%} = \frac{42,154 \text{ lb}}{\text{d}}$$

7.
$$\frac{24.6 \text{ L}}{1 \text{ s}} \times \frac{60 \text{ s}}{1 \text{ min}} \times \frac{60 \text{ min}}{1 \text{ hr}} \times \frac{24 \text{ hr}}{1 \text{ d}} \times \frac{1 \text{ kg}}{1 \text{ L}} \times \frac{0.9\%}{100.0\%}$$

$$= \frac{19,129 \text{ kg}}{\text{d}}$$

8. 52 mg/L were left in the effluent so the amount removed was:

$$135 \text{ mg/L} - 52 \text{ mg/L} = 83 \text{ mg/L}$$

The percentage must be based on what you started with (135 mg/L)

$$\% \text{ Removal} = \frac{83 \text{ mg/L}}{135 \text{ mg/L}} \times 100\% = 61.5\%$$

9. Amount Removed $= 14.5 \text{ mL/L} - 1.3 \text{ mL/L} = 13.2 \text{ mL/L}$

$$\% \text{ Removed} = \frac{13.2 \text{ mL/L}}{14.5 \text{ mL/L}} \times 100\% = 91.0\%$$

10.

$$\text{WOR} = \frac{1,380,000 \text{ gpd}}{70 \text{ ft}} = 19,714 \text{ gpd/ft}$$

11.

$$\text{WOR} = \frac{5175 \text{ m}^3/\text{d}}{21.5 \text{ m}} = 240.7 \text{ m}^3/\text{d/m}$$

12. a.
$$\text{Tank Volume} = 82 \text{ ft} \times 21 \text{ ft} \times \times 13 \text{ ft} \times \frac{7.48 \text{ gal}}{1 \text{ ft}^3}$$

$$= 167,447 \text{ gal}$$

Detention Time $= 3 \text{ tanks} \times 167,447 \text{ gal}$

$$\times \frac{1 \text{ d}}{5,100,000 \text{ gal}} \times \frac{24 \text{ hr}}{1 \text{ d}} \times \frac{60 \text{ min}}{1 \text{ hr}} = 142 \text{ min}$$

$$\text{SOR} = \frac{5,100,000 \text{ gpd}}{3 \times 82 \text{ ft} \times 21 \text{ ft}} = \frac{987 \text{ gpd}}{\text{ft}^2}$$

$$\text{WOR} = \frac{5,100,000 \text{ gpd}}{3 \times 88 \text{ ft}} = \frac{19,318 \text{ gpd}}{\text{ft}}$$

b. Of the three constraints given, only the weir overflow rate has but a single value to satisfy. Determine the number of tanks to satisfy the weir overflow rate and check if that number also falls in the ranges for detention time and surface overflow rate.

$$\text{WOR} = \frac{9,800,000 \text{ gpd}}{N \text{ tanks} \times 88 \text{ ft}} = \frac{20,000 \text{ gpd}}{\text{ft}}$$

Rearrange to get N by itself. Multiply both sides by N. Divide both sides by 20,000.

$$\frac{9,800,000 \text{ gpd}}{N \text{ tanks} \times 88 \text{ ft}} \times \frac{N}{20,000} = 20,000 \text{ gpd/ft} \times \frac{N}{20,000}$$

The Ns cancel on the left side, the 20,000s cancel on the right.

$$\frac{9,800,000 \text{ gpd}}{88 \text{ ft} \times 20,000 \text{ gpd/ft}} = N \text{ tanks} = 5.5 \text{ tanks} \rightarrow 6 \text{ tanks}$$

Double check the weir overflow rate with six tanks:

$$\text{WOR} = \frac{9,800,000 \text{ gpd}}{6 \times 88 \text{ ft}} = \frac{18,560 \text{ gpd}}{\text{ft}} \text{ (Less than 20,000 gpd/ft)}$$

Check the detention time and surface overflow rate for six tanks:

$$\text{Detention Time} = 6 \text{ tanks} \times 167,447 \text{ gal}$$

$$\times \frac{1 \text{ d}}{9,800,000 \text{ gal}} \times \frac{24 \text{ hr}}{1 \text{ d}} \times \frac{60 \text{ min}}{1 \text{ hr}}$$

$$= 147.6 \text{ min (Between 90 and 150 min)}$$

$$\text{SOR} = \frac{9,800,000 \text{ gpd}}{6 \times 82 \text{ ft} \times 21 \text{ ft}}$$

$$= \frac{948 \text{ gpd}}{\text{ft}^2} \text{ (Between 800 and 1000 gpd/ft}^{2)}$$

Operating six tanks satisfies all three requirements. Does seven?

---

13. a. $\text{Tank Volume} = 25 \text{ m} \times 6.4 \text{ m} \times 4 \text{ m} = 640 \text{ m}^3$

$$\text{Detention Time} = 3 \text{ tanks} \times 640 \text{ m}^3 \times \frac{1 \text{ d}}{19,300 \text{ m}^3} \times \frac{24 \text{ hr}}{1 \text{ d}}$$

$$\times \frac{60 \text{ min}}{1 \text{ hr}} = 143 \text{ min}$$

$$\text{SOR} = \frac{19,300 \text{ m}^3/\text{d}}{3 \times 25 \text{ m} \times 6.4 \text{ m}} = \frac{40.2 \text{ m}^3/\text{d}}{\text{m}^2}$$

$$\text{WOR} = \frac{19{,}300 \text{ m}^3/\text{d}}{3 \times 26.8 \text{ m}} = \frac{240 \text{ m}^3/\text{d}}{\text{m}}$$

b. Of the three constraints given, only the weir overflow rate has but a single value to satisfy. Determine the number of tanks to satisfy the weir overflow rate and check if that number also falls in the ranges for detention time and surface overflow rate.

$$\text{WOR} = \frac{37{,}100 \text{ m}^3/\text{d}}{N \text{ tanks} \times 26.8 \text{ m}} = \frac{250 \text{ m}^3/\text{d}}{\text{m}}$$

Rearrange to N by itself. Multiply both sides by N. Divide both sides by 250.

$$\frac{37{,}100 \text{ m}^3/\text{d}}{N \text{ tanks} \times 26.8 \text{ m}} \times \frac{N}{250} = \frac{250 \text{ m}^3/\text{d}}{\text{m}} \times \frac{N}{250}$$

The Ns cancel on the left side, the 250s cancel on the right.

$$\frac{37{,}100 \text{ m}^3/\text{d}}{26.8 \text{ m} \times 250 \text{ m}^3/\text{d}/\text{m}} = N \text{ tanks} = 5.5 \text{ tanks} => 6 \text{ tanks}$$

Double check the weir overflow rate with six tanks:

$$\text{WOR} = \frac{37{,}100 \text{ m}^3/\text{d}}{6 \times 26.8 \text{ m}} = \frac{230.7 \text{ m}^3/\text{d}/\text{m}}{\text{m}} \quad (\text{Less than } 25 \text{ m}^3/\text{d}/\text{m})$$

Check the detention time and surface overflow rate for six tanks:

$$\text{Detention Time} = 6 \text{ tanks} \times 640 \text{ m}^3 \times \frac{1 \text{ d}}{37{,}100 \text{ m}^3} \times \frac{24 \text{ hr}}{1 \text{ d}}$$

$$\times \frac{60 \text{ min}}{1 \text{ hr}} = 149 \text{ min} \ (\text{Between } 90 \text{ and } 150 \text{ min})$$

$$\text{SOR} = \frac{37{,}100 \text{ m}^3/\text{d}}{6 \times 25 \text{ m} \times 6.4 \text{ m}} = \frac{38.6 \text{ m}^3/\text{d}}{\text{m}^2}$$

(Between 32 and 41 m³/d/m²)

Operating six tanks satisfies all three requirements. Does seven?

ANSWERS TO THE QUESTIONS: TRICKLING FILTERS

1. The slime that grows on the trickling filter media is called the zoogleal mass.

2. Bugs grow on the media and eat the food as it passed by them. However, the water by no means is strained or filtered as the name of the unit might indicate.

3. The process in which the slime layer gets too thick and falls off is referred to as sloughing.

4. Hydraulic loading rate is based on the application of flow across the top surface of the filter. The filter volume is of no concern in this particular calculation.

5. The recirculation ratio is the recirculated flowrate divided by the primary effluent flowrate.

6. The recirculation flowrate can be controlled by the dissolved oxygen or DO of the flow leaving the filter. If the DO drops below 1.5 mg/L the recirculation can be increased. If the DO goes above 2.0 mg/L the recirculation can be decreased.

7. Unclarified trickling filter effluent contains the sloughings from the unit. If the unit is just starting up or if something killed some of the slime bugs in the zoogleal mass, the unit can be restocked or reseeded by recirculating this unclarified effluent.

8. The organic loading rate is concerned with the total media surface that the bugs can grow on to eat the food. For this reason, the entire volume of the unit is used in this calculation.

9. The two major categories of synthetic media are corrugated plastic sheets and randomly placed individual media pieces.

10. Only about 40% of rock media is open space.

11. Plastic media contains about 95% open space.

12. Open space in a trickling filter allows the water to pass through the filter, the sloughings to pass down through and be carried out of the filter and for air to circulate through the filter.

13. A trickling filter with a media bed greater than 10 feet deep is sometimes referred to as a biotower.

14. The four categories of trickling filters based on their organic loading rate are Low Rate, Intermediate Rate, High Rate, and Roughing.

15. High rate plastic media has much more surface area that must be kept wetted. For this reason a hydraulic loading rate of 1000 gpd/ft^2 or 40 m^3/d/m^2 is recommended for all organic loading rates.

16. Ponding can be described as water seen standing on the top of the trickling filter.

17. During winter operation, as recirculation is increased the water will get colder.

18. The worms and snails may play the role of breaking up slime mass to help with sloughing.

19. If a trickling filter bed gets too cold, the worms and snails will move lower in the filter for warmth.

20. To monitor even filter distribution a line of pans or similar containers can be kept from the filter center to the outer edge. Even distribution is determined by an even filling of the pans.

21. Some facilities have controlled filamentous growth in their trickling filter through the application of activated sludge.

SOLUTIONS TO THE PROBLEMS: TRICKLING FILTERS

1.
$$0.296 \text{ MGD} + 0.348 \text{ MGD} = 0.644 \text{ MGD}$$

$$\text{Surface Area} = 0.785 \times (80\text{ft})^2$$

$$= 5024 \text{ ft}^2$$

$$\frac{644{,}000 \text{ gpd}}{5024 \text{ ft}^2} = 128 \text{ gpd}/\text{ft}^2$$

2.
$$1046 \text{ m}^3/\text{d} + 1130 \text{ m}^3/\text{d} = 2176 \text{ m}^3/\text{d}$$

$$\text{Surface Area} = 0.785 \times (25\text{m})^2$$

$$= 490.6 \text{ m}^2$$

$$\frac{2176 \text{ m}^3/\text{d}}{490.6 \text{ m}^2} = 4.4 \text{ m}^3/\text{d}/\text{m}^2$$

3.
$$\frac{4.25 \text{ MGD}}{2.50 \text{ MGD}} = 1.7$$

4.
$$\frac{16{,}000 \text{ m}^3/\text{d}}{9500 \text{ m}^3/\text{d}} = 1.7$$

5. a.
$$74 \text{ mg/L} \times 1.15 \text{ MGD} \times \frac{8.34 \text{ lb}}{1 \text{ gal}} = \frac{709.7 \text{ lb}}{\text{d}}$$

$$\text{Surface Area} = 0.785 \times (70 \text{ ft})^2$$

$$= 3846.5 \text{ ft}^2$$

$$\text{Surface Area} \times \text{Depth} = \text{Volume}$$

$$3846.5 \text{ ft}^2 \times 5 \text{ ft} = 19{,}232.5 \text{ ft}^3$$

$$\frac{709.7 \text{ lb BOD/d}}{19{,}232.5 \text{ ft}^3} \times \frac{1000 \text{ (number)}}{1000 \text{ (unit)}}$$

$$= \frac{709.7 \times 1000}{19{,}232{,}5} \frac{\text{lb BOD/d}}{1000 \text{ ft}^3}$$

$$= 36.9 \, \frac{\text{lb BOD/d}}{1000 \, \text{ft}^3}$$

b. What is an acre-foot? It's just another measure of volume. Just as 1 cubic foot (1 ft³) is the volume in a box 1 ft by 1 ft by 1 ft, an acre-foot is the volume in a box with a bottom 1 acre in size and a depth of 1 foot. You can fit 43,560 cubic feet inside 1 acre-foot (ac-ft). The box doesn't necessarily have to be 1 acre in size. Acre-feet is just another way to talk about volume, regardless of the specific size or shape of the volume.

Conversion Factor: 43,560 ft³ = 1 ac-ft

$$19,232.5 \, \text{ft}^3 \times \frac{1 \, \text{ac-ft}}{43,560 \, \text{ft}^3} = 0.4415 \, \text{ac-ft}$$

$$\frac{709.7 \, \text{lbs BOD/d}}{0.4415 \, \text{ac-ft}} = \frac{1607 \, \text{lbs BOD/d}}{\text{ac-ft}}$$

6.

$$\frac{4350 \, \text{m}^3/\text{d} \times 74 \, \text{mg/L}}{1000} = 321.9 \, \text{kg BOD/d}$$

$$\text{Surface Area} = 0.785 \times (21 \, \text{m})^2$$

$$= 346.2 \, \text{m}^2$$

$$346.2 \, \text{m}^2 \times 1.5 \, \text{m} = 519.3 \, \text{m}^3$$

$$= \frac{321.9 \, \text{kg BOD/d}}{519.3 \, \text{m}^3} \times \frac{100 \, (\text{number})}{100 \, (\text{unit})}$$

$$= 62 \, \frac{\text{kg BOD/d}}{100 \, \text{m}^3}$$

7.

$$79 \, \text{mg/L} - 14 \, \text{mg/L} = 65 \, \text{mg/L removed}$$

$$65 \, \text{mg/L} \times 3.7 \, \text{MGD} \times \frac{8.34 \, \text{lb}}{\text{gal}} = 2006 \, \text{lb BOD/day}$$

8.

$$79 \, \text{mg/L} - 14 \, \text{mg/L} = 65 \, \text{mg/L removed}$$

$$\frac{14,000 \, \text{m}^3/\text{d} \times 65 \, \text{mg/L}}{1000} = 910 \, \text{kg BOD/d}$$

9.   Surface Area $= 0.785 \times (25\ \text{ft})^2$

$$= 490.6\ \text{ft}^2$$

Volume $= 490.6\ \text{ft}^2 \times 12\ \text{ft}$

$$= 5887.2\ \text{ft}^3$$

Organic Loading Applied

$$\frac{24\ \text{lb BOD/d}}{1000\ \text{ft}^3} \times 5887.2\ \text{ft}^3 \times \frac{1000\ (\text{unit})}{1000\ (\text{number})}$$

$$= \frac{24 \times 5887.2}{1000}\ \text{lb BOD/d}$$

$$= 141.3\ \text{lb BOD/d}$$

$141.3\ \text{lb BOD/d} = 80\ \text{mg/L} \times (\text{Primary Eff})\ \text{MGD}$

$$\times\ 8.34\ \text{lb/gal}$$

$$\frac{141.3\ \text{lb BOD/d}}{80\ \text{mg/l} \times 8.34\ \text{lbs/gal}} = (\text{PE})\ \text{MGD}$$

$$= 0.212\ \text{MGD}$$

Hydraulic loading rate for synthetic media should be kept at 1000 gpd/ft².

$$\frac{1000\ \text{gpd}}{\text{ft}^2} = \frac{(\text{Applied})\ \text{gpd}}{490.6\ \text{ft}^2}$$

$$\frac{1000\ \text{gpd}}{\text{ft}^2} \times 490\ \text{ft}^2 = (\text{Applied})\ \text{gpd}$$

$$= 490,600\ \text{gpd}$$

Applied Flowrate $=$ Primary Effluent Flowrate

$$+\ \text{Recirculated Flowrate}$$

Applied $-$ Primary Effluent $=$ Recirculated

$$490,600 \text{ gpd} - 212,000 \text{ gpd} = 278,600 \text{ gpd (recirculated)}$$

$$\frac{278,600 \text{ gal}}{d} \times \frac{1 \text{ d}}{1440 \text{ min}} = 193.5 \text{ gpm}$$

10. $$\text{Surface Area} = 0.785 \times (7.6 \text{ m})^2$$

$$= 45.34 \text{ m}^2$$

$$\text{Volume} = 45.34 \text{ m}^2 \times 3.6 \text{ m}$$

$$= 163.22 \text{ m}^3$$

Organic Loading Applied

$$\frac{39 \text{ kg BOD/d}}{100 \text{ m}^3} \times 163.22 \text{ m}^3 \times \frac{100 \text{ (unit)}}{100 \text{ (number)}}$$

$$= \frac{39 \times 163.22}{100} \text{ kg BOD/d}$$

$$= 63.66 \text{ kg BOD/d}$$

$$63.66 \text{ kg BOD/d} = \frac{80 \text{ mg/L} \times (\text{Primary Eff}) \text{ m}^3/d}{1000}$$

$$\frac{63.66 \text{ kg BOD/d} \times 1000}{80 \text{ mg/L}} = (\text{PE}) \text{ m}^3/d$$

$$= 795.7 \text{ m}^3/d$$

Hydraulic loading rate for synthetic media should be kept at 40 $m^3/d/m^2$.

$$\frac{40 \text{ m}^3/d}{m^3} = \frac{(\text{Applied}) \text{ m}^3/d}{45.34 \text{ m}^2}$$

$$\frac{40 \text{ m}^3/d}{m^2} \times 45.34 \text{ m}^2 = (\text{Applied}) \text{ m}^3/d$$

$$= 1813.6 \text{ m}^3/d$$

Applied Flowrate = Primary Effluent Flowrate

+ Recirculated Flowrate

Applied − Primary Effluent = Recirculated

1813.6 m³/d − 795.7 m³/d = 1017.9 m³/d (recirculated)

$$\frac{1017.9 \text{ m}^3}{\text{d}} \times \frac{1 \text{ d}}{1440 \text{ min}} \times \frac{1 \text{ min}}{60 \text{ s}} \times \frac{1000 \text{ L}}{1 \text{ m}^3} = \frac{11.8 \text{ L}}{\text{s}}$$

11. (1) Determine the surface area.

Surface area per tank = 0.785 (20 m)²

= 314 m²

2 tanks are in service ("Both . . . filters in service . . .")

314 m² × 2 tanks = 628 m²

(2) Since the hydraulic loading is in units of L/s, convert the recirculated flowrate.

$$\frac{5000 \text{ m}^3}{\text{d}} \times \frac{1000 \text{ L}}{1 \text{ m}^3} \times \frac{1 \text{ d}}{1440 \text{ min}} \times \frac{1 \text{ min}}{60 \text{ s}} = \frac{57.9 \text{ L}}{\text{s}}$$

(3) The hydraulic loading rate is the total flow applied (influent + recirculated) divided by the surface area. Solve for the influent flowrate.

$$\frac{\text{HLR L/s}}{\text{m}^2} = \frac{\text{Influent L/s} + \text{Recirc. L/s}}{\text{Surface Area m}^2}$$

$$\frac{0.3 \text{ L/s}}{\text{m}^2} = \frac{\text{Influent L/s} + 57.9 \text{ L/s}}{628 \text{ m}^2}$$

$$\frac{0.3 \text{ L/s} \times 628 \text{ m}^2}{\text{m}^2} = \text{Influent L/s} + 57.9 \text{ L/s}$$

188.4 L/s = Influent L/s + 57.9 L/s

$$188.4 \text{ L/s} - 57.9 \text{ L/s} = \text{Influent L/s}$$

$$= 130.5 \text{ L/s}$$

(4) From the influent flowrate and the desired hydraulic loading the new recirculated flowrate can be calculated.

$$\frac{0.25 \text{ L/s}}{m^2} = \frac{130.5 \text{ L/s} + \text{Recirc. L/s}}{628 \text{ m}^2}$$

$$\frac{(0.25 \text{ L/s} \times 628 \text{ m}^2)}{m^2} - 130.5 \text{ L/s} = \text{Recirc. L/s}$$

$$= 26.5 \text{ L/s}$$

(5) Determine the difference in the recirculation rates and convert to m^3/d.

$$57.9 \text{ L/s} - 26.5 \text{ L/s} = 31.4 \text{ L/s}$$

$$\frac{31.4 \text{ L}}{s} \times \frac{60 \text{ s}}{1 \text{ min}} \times \frac{1440 \text{ min}}{1 \text{ d}} \times \frac{1 \text{ m}^3}{1000 \text{ L}} = 2713 \text{ m}^3/d$$

The recirculated flowrate must be lowered by 2713 m^3/d in order to get the desired hydraulic loading rate.

ANSWERS TO THE QUESTIONS: ROTATING BIOLOGICAL CONTACTORS

1. The purpose of an RBC is to remove soluble food from the wastewater.

2. An RBC must follow primary settling in order to remove the settleable solids. An RBC is designed to treat only soluble material.

3. About 40% of the RBC disk is submerged.

4. RBC media disks are 12 ft or 3.6 m in diameter.

5. An RBC shaft is 25 ft or 7.5 m long.

6. A standard media shaft provides 100,000 ft² or 9300 m² of surface area.

7. RBCs are typically operated at a rotational speed between 1.5 and 1.6 rpm.

8. An RBC is followed by a settling tank to remove sloughed solids.

9. An RBC should be covered to: (1) prevent biomass from freezing, (2) prevent rain from washing the biomass off, (3) prevent algae growth, and (4) prevent sun damage to the media.

10. An RBC stage is considered the media rotated in a tank area not separated by a baffle.

11. A healthy first stage biomass is uniform, thin, and light brown.

197

12. White biomass indicated filamentous bacteria growth.

13. Filamentous bacteria eat sulfur chemicals.

14. Sodium Nitrate ($NaNo_3$) can be added to increase oxygen concentration.

15. The 1st stage DO should be no less than 0.5 mg/L.

16. The last stage DO should be no less than 2.0 mg/L.

17. The hydraulic loading to an RBC should be between 1 and 3 gpd/ft² or between 0.04 and 0.12 m³/d/m².

18. An organically overloaded first stage biomass will appear heavy and shaggy. If the problem is not corrected, anaerobic conditions may develop, which may be followed by white patches of filamentous bacteria in the biomass.

19. The other part of the total BOD is the particulate BOD.

20. In the soluble BOD test the sample is filtered before doing the same procedure as for a total BOD test.

21. The particulate BOD can be calculated by the formula:

$$\text{Particulate BOD} = k \times \text{SS}$$

22. The value of k will range from 0.5 to 0.7 for domestic sewage.

23. The organic loading based on SBOD that an RBC usually receives is from 3 to 4 lb/d/1000 ft² or from 1.5 to 2.0 kg/d/100 m².

24. The organic loading that a first stage of an RBC usually receives is from 4 to 6 lb/d/1000 ft² or from 2.0 to 2.9 kg/d/100 m².

25. The two types of RBC drive units are the mechanical drive and the air drive.

26. The three advantages of air drives are: (1) less torque applied to the shaft, (2) thinner biomass, and (3) higher DOs with reduced filamentous growth.

27. The three disadvantages of air drives are: (1) unbalanced or stopped rotation, (2) more overall power requirements, and (3) air diffuser adjustment required.

28. RBC media is made of sheets of polyethylene.

29. A standard density media shaft provides 100,000 ft² or 9300 m² of surface area.

30. A third stage biomass is thinner than a first stage biomass.

31. Two types of RBC media are Standard Density and High Density.

32. High density media has more available surface area because the polyethylene sheets are bonded closer together.

33. If high density media was installed in the first stage the biomass would grow too thick and close off the spacings between sheets.

34. A high density media shaft provides from 120,000 to 180,000 ft² or from 11,200 to 16,700 m² of surface area.

35. Snails cause problems in a final RBC stage because they eat the bugs needed in the process.

36. The length of the snail's life cycle is very much dependent on the wastewater temperature.

37. Copper sulfate has been used in three doses to effectively remove snails.

SOLUTIONS TO THE PROBLEMS: ROTATING BIOLOGICAL CONTACTORS

1.
$$\frac{350,000 \text{ gpd}}{195,000 \text{ ft}^2} = 1.8 \text{ gpd/ft}^2$$

2.
$$\frac{1320 \text{ m}^3/\text{d}}{18,100 \text{ m}^2} = 0.073 \text{ m}^3/\text{d/m}^2$$

3.
$$\text{TBOD} = \text{SBOD} + (k \times \text{SS})$$

$$\text{TBOD} - (k \times \text{SS}) = \text{SBOD}$$

$$180 \text{ mg/L} - (0.5 \times 200 \text{ mg/L}) = \text{SBOD}$$

$$= 80 \text{ mg/L}$$

---

4. $$\frac{1.4 \text{ MGD} \times 122 \text{ mg/L} \times 8.34 \text{ lb/gal}}{392{,}000 \text{ ft}^2} \times \frac{1000 \text{ (number)}}{1000 \text{ (unit)}}$$

$$= 3.6 \text{ lb/d/1000 ft}^2$$

---

5. $$\frac{5300 \text{ m}^3/\text{d} \times 122 \text{ mg/L}}{1000} = 646.6 \text{ kg/d}$$

$$\frac{646.6 \text{ kg/d}}{36{,}400 \text{ m}^2} \times \frac{100 \text{ (number)}}{100 \text{ (unit)}} = 1.8 \text{ kg/d/100 m}^2$$

---

6. $$\text{SBOD} = \text{TBOD} - (k \times \text{SS})$$

$$= 202 \text{ mg/L} - (0.65 \times 225 \text{ mg/L})$$

$$= 56 \text{ mg/L}$$

$$\frac{0.69 \text{ MGD} \times 56 \text{ mg/L} \times 8.34 \text{ lb/gal}}{104{,}000 \text{ ft}^2} \times \frac{1000 \text{ (number)}}{1000 \text{ (unit)}}$$

$$= \frac{3.1 \text{ lb/d}}{1000 \text{ ft}^2}$$

---

7. $$\text{SBOD} = \text{TBOD} - (k \times \text{SS})$$

$$= 202 \text{ mg/L} - (0.65 \times 225 \text{ mg/L})$$

$$= 56 \text{ mg/L}$$

$$\frac{2610 \text{ m}^3/\text{d} \times 56 \text{ mg/L}}{1000} = 146.2 \text{ kg/d}$$

$$\frac{146.2 \text{ kg/d}}{9660 \text{ m}^2} \times \frac{100 \text{ (number)}}{100 \text{ (unit)}} = 1.5 \text{ kg/d/100 m}^2$$

---

8. a. $$\frac{445{,}000 \text{ gpd}}{204{,}000 \text{ ft}^2} = 2.2 \text{ gpd/ft}^2$$

b.
$$SBOD = TBOD - (k \times SS)$$

$$= 237 \text{ mg/L} - (0.5 \times 148)$$

$$= 163 \text{ mg/L}$$

$$0.445 \text{ MGD} \times 163 \text{ mg/L} \times 8.34 \text{ lb/gal} = 604.9 \text{ lb/d}$$

$$\frac{604.9 \text{ lb/d}}{204,000 \text{ ft}^2} \times \frac{1000 \text{ (number)}}{1000 \text{ (unit)}} = 3.0 \text{ lb/d/1000 ft}^2$$

c.
$$\frac{604.9 \text{ lb/d}}{(0.75 \times 102,000) \text{ ft}^2} \times \frac{1000 \text{ (number)}}{1000 \text{ (unit)}} = 7.9 \text{ lb/d/1000 ft}^2$$

The hydraulic loading and unit organic loading are in the proper ranges but the first stage organic loading is too high. If the size of the first stage is increased by 25% to include one entire shaft, the first stage organic loading is calculated to be:

$$\frac{604.9 \text{ lb/d}}{102,000 \text{ ft}^2} \times \frac{1000 \text{ (number)}}{1000 \text{ (unit)}} = 5.9 \text{ lb/d/1000 ft}^2$$

This process change would now bring the first stage organic loading into an acceptable range.

---

9. a.
$$\frac{1714 \text{ m}^3/\text{d}}{(2 \times 9475) \text{ m}^2} = 0.090 \text{ m}^3/\text{d/m}^2$$

b.
$$SBOD = TBOD - (k \times SS)$$

$$= 237 \text{ mg/L} - (0.5 \times 148)$$

$$= 163 \text{ mg/L}$$

$$\frac{1714 \text{ m}^3/\text{d} \times 163 \text{ mg/L}}{1000} = 279.4 \text{ kg/d}$$

$$\frac{279.4 \text{ kg/d}}{(2 \times 9475) \text{ m}^2} \times \frac{100 \text{ (number)}}{100 \text{ (unit)}} = 1.5 \text{ kg/d/m}^2$$

c.
$$\frac{279.4 \text{ kg/d}}{(0.75 \times 9475) \text{ m}^2} \times \frac{100 \text{ (number)}}{100 \text{ (unit)}} = 3.9 \text{ kg/d/m}^2$$

The hydraulic loading and unit organic loading are in the proper ranges but the first stage organic loading is too high. If the size of the first stage is increased by 25 % to include one entire shaft, the first stage organic loading is calculated to be:

$$\frac{279.4 \text{ kg/d}}{9475 \text{ m}^2} \times \frac{100 \text{ (number)}}{100 \text{ (unit)}} = 2.9 \text{ kg/d/m}^2$$

This process change would now bring the first stage organic loading into an acceptable range.

ANSWERS TO THE QUESTIONS: ACTIVATED SLUDGE

1. The air being put into an aeration tank provides the needed oxygen for the bugs and the required mixing to put the food and bugs together.

2. The watery mixture of bugs and solids removed from the bottom of the settling tank is called activated sludge.

3. The primary effluent contains the food and the return sludge contains the bugs.

4. The mixture of primary effluent and return sludge is called mixed liquor.

5. The activated sludge removed from the bottom of the clarifier is split into return activated sludge and waste activated sludge.

6. The major purpose for an activated sludge unit is to remove BOD.

7. The official name for the little bugs is bacteria.

8. The kind of little bugs that eat things like spaghetti sauce and other carbon-containing food are called carbonaceous bacteria.

9. The kind of little bugs that eat nitrogen-containing chemicals are called nitrogenous bacteria.

10. The small bugs stick together by the fat coating on the outside of their bodies.

11. The big bug that might be called the ''blob'' is the amoeba because it has no definite shape.

12. The big bug that uses a whip to move around is called a flagellate.

13. The first type of ''hairy'' big bug to show up after a new plant start-up is the free-swimming ciliate.

14. The two kinds of big bugs that the operator hopes to have in greatest number are the crawling and stalked ciliates.

15. An operator needs some filaments in the sludge to provide the necessary framework for the floc particles.

16. If the sludge has too many filaments in it, the sludge will be slow in compacting, which is referred to as bulking.

17. The flow that is always put into an aeration tank first is the return sludge.

18. A plug-flow aeration tank has the primary effluent and return sludge added at the beginning of the tank and they pass together through a long and narrow aeration tank often folded into three or four passes.

19. One advantage to using a plug-flow aeration tank is that it prevents short circuiting. One disadvantage is that it has a high oxygen demand at the beginning of the first pass.

20. Step aeration means that more air is added to the first pass and less to each of the rest. Step feed means the primary effluent is added to the mixed liquor at more than one location.

21. The air distribution in a four-pass tapered aeration tank might be set up with 40% of the diffusers in the first pass, 25% in the second, 20% in the third, leaving 15% of the diffusers in the final pass.

22. At least twenty minutes contact time should be allowed for the bugs to get their food.

23. The bugs need about two hours stabilization time to digest their food.

24. Some advantages to using contact stabilization include less tank space required, less air required, and a reserve supply of bugs to help handle hydraulic and toxic shock loads.

25. Step feed can reduce oxygen requirements by giving less food to the bugs at one location, which reduces their oxygen demand at that location.

26. Changing from big-bubble to small-bubble diffusers will reduce air requirements because the transfer of oxygen into the water will be more efficient.

27. The three main things an operator can control when operating an activated sludge unit are the aeration rate, the return sludge rate, and the waste sludge rate.

28. Filamentous bugs might grow in larger numbers if there isn't enough air put into the aeration tank.

29. A safe minimum dissolved oxygen concentration to stay above when operating an activated sludge unit is 1.5 mg/L.

30. The name given to the sludge settled on the bottom of an FST is the sludge blanket.

31. An operator can measure the depth of the sludge blanket in a clarifier using a core sampler, sometimes also called a sludge judge.

32. A good sludge blanket thickness to keep is about 2 feet.

33. A settlometer test shows the operator how the sludge is settling in the clarifier.

34. The best size settlometer to use is the 2 liter size because it will have less interference from the side walls as the sludge settles.

35. The disadvantage to using a constant return pumping rate is that there will be a constant daily shifting of solids between aeration and the settling tanks.

36. A target MLSS would be fine to use only if the plant flowrates and BOD loadings are fairly consistent.

37. The maximum percentage the wasting rate should be changed in one day is 10%.

38. The target MLSS concentration will be lowered for summer operation because the bugs work faster in warmer temperatures, so fewer of them are needed.

39. There is no difference between MCRT and SRT. In the current wastewater literature these two terms are used interchangeably.

40. The sludge age is based on the bugs coming into the activated sludge unit.

41. The two paths that the bugs can take to leave an activated sludge unit are the waste sludge and the clarified effluent.

42. The bugs in the return sludge should not be used in an SRT calculation because the return sludge is kept inside the system.

43. The volatile suspended solids are measured by heating the dried solids to 500 °C to allow the volatiles to ''burn'' off. Reweigh after cooling and the weight difference is the volatile portion.

44. The operator gets *F/M* data in less time by using COD data to estimate the BOD data or just use the COD as the food directly.

45. Variations in the *F/M* data can be handled by using a running seven-day average to base wasting rate adjustments.

46. A magnification of 400 times (400×) should work well.

47. Add a drop of 1% nickel sulfate to slow the bugs down.

48. The two big bugs that would be best to have in greatest number in the sludge are the crawling and stalked ciliates.

49. The method that makes use of a lab centrifuge is called sludge quality.

50. If nocardia bugs are causing a foam problem there will be 10 to 100 times more of them in the foam compared to those in the mixed liquor.

51. No, nocardia bugs do not cause problems with sludge settling because their filaments are two short.

52. The two properties that make nocardia bugs such a problem in the foam is they have a waxy coating that allows them to float and their filaments interlock.

53. Plant conditions that favor the breeding of nocardia bugs include high influent grease concentration, poor PST scum removal, and recycle of scum without a removal location.

54. The dead nocardia bugs add to the problem because they continue to float and accumulate with the live bugs in the foam.

55. Nocardia foam may need to be removed from the aeration tanks by vacuum truck.

56. Bulking sludge is sludge that does settle but compacts very slowly.

57. To get an early indication of potential sludge bulking, the sludge should be examined by microscope at least once per week.

58. If the return sludge is chlorinated, chlorine should be dosed at a rate of $1-10$ pounds of chlorine per 1000 pounds of mixed liquor volatile suspended solids.

59. If chlorination of the return sludge is used, the sludge settleability should improve in 1 to 3 days.

60. Some temporary remedies for a sludge bulking problem include increasing return sludge rates, dosing polymer to aid with settling, and chlorinating the return sludge.

SOLUTIONS TO THE PROBLEMS: ACTIVATED SLUDGE

1.
$$\frac{R}{Q} = \frac{430 \text{ mL}}{2000 \text{ mL} - 430 \text{ mL}} = \frac{430 \text{ mL}}{1570 \text{ mL}} = 0.274$$

$$\frac{R}{Q} \times \frac{Q}{1} = 0.274 \times 5.9 \text{ MGD} = 1.62 \text{ MGD}$$

$$\frac{1.62 \text{ million gal}}{1 \text{ d}} \times \frac{1 \text{ d}}{1440 \text{ min}} \times \frac{1,000,000 \text{ gal}}{1 \text{ million gal}}$$

$$= 1125 \text{ gpm}$$

2.

$$\frac{R}{Q} = \frac{430 \text{ mL}}{2000 \text{ mL} - 430 \text{ mL}} = \frac{430 \text{ mL}}{1570 \text{ mL}} = 0.274$$

$$\frac{R}{Q} \times \frac{Q}{1} = 0.274 \times 22{,}300 \text{ m}^3/\text{d} = 6110 \text{ m}^3/\text{d}$$

3. We have: MLSS = 2200 mg/L

   We want: MLSS = 2150 mg/L

   $\overline{ 50 \text{ mg/L}}$ (in excess)

   $$50 \text{ mg/L} \times 0.42 \text{ MG} \times 8.34 \text{ lbs/gal} = 175 \text{ pounds}$$

   $$175 \text{ lb/d} = 4950 \text{ mg/L} \times \text{Additional WAS (MGD)} \times 8.34 \text{ lb/gal}$$

   $$\frac{175}{4950 \times 8.34} = 0.004 \text{ MGD (Additional WAS)}$$

   New WAS = 0.118 MGD + 0.004 MGD = 0.122 MGD

4. We have: MLSS = 2200 mg/L

   We want: MLSS = 2150 mg/L

   $\overline{ 50 \text{ mg/L}}$ (in excess)

   $$\frac{50 \text{ mg/L} \times 1590 \text{ m}^3}{1000} = 79.5 \text{ kg}$$

   $$79.5 \text{ kg/d} = \frac{4950 \text{ mg/L} \times \text{Additional WAS (m}^3/\text{d)}}{1000}$$

   $$\frac{79.5 \times 1000}{4950} = 16 \text{ m}^3/\text{d (Additional WAS)}$$

   New WAS = 447 m³/d + 16 m³/d = 463 m³/d

5. We have: MLSS = 2080 mg/L

   We want: MLSS = 2150 mg/L

   $\overline{ -70 \text{ mg/L}}$

   We need an extra 70 mg/L MLSS in aeration.

   $$70 \text{ mg/L} \times 0.42 \text{ MG} \times 8.34 \text{ lb/gal} = 245 \text{ pounds (needed)}$$

245 lb/day = 4890 mg/L × Excess WAS (MGD) × 8.34 lb/gal

$$\frac{245}{4890 \times 8.34} = 0.006 \text{ MGD (Excess WAS)}$$

$$WAS = \frac{86.1 \text{ gal}}{1 \text{ min}} \times \frac{1440 \text{ min}}{1 \text{ d}} \times \frac{1 \text{ million gal}}{1{,}000{,}000 \text{ gal}}$$

$$= 0.124 \text{ MGD}$$

New WAS = 0.124 MGD − 0.006 MGD = 0.118 MGD

$$\frac{0.118 \text{ million gal}}{1 \text{ d}} \times \frac{1 \text{ d}}{1440 \text{ min}} \times \frac{1{,}000{,}000 \text{ gal}}{1 \text{ million gal}}$$

$$= 81.9 \text{ gpm}$$

---

6. We have: MLSS = 2080 mg/L

We want: MLSS = 2150 mg/L
$$\frac{}{-70 \text{ mg/L}}$$

We need an extra 70 mg/L MLSS in aeration.

$$\frac{70 \text{ mg/L} \times 1590 \text{ m}^3}{1000} = 111 \text{ kg (needed)}$$

$$111 \text{ kg/d} = \frac{4890 \text{ mg/L} \times \text{Excess WAS (m}^3/\text{d)}}{1000}$$

$$\frac{111 \times 1000}{4890} = 23 \text{ m}^3/\text{d (Excess WAS)}$$

$$WAS = \frac{5.43 \text{ L}}{1} \times \frac{60 \text{ s}}{1 \text{ min}} \times \frac{1440 \text{ min}}{1 \text{ d}} \times \frac{1 \text{ m}^3}{1000 \text{ L}} = 469 \text{ m}^3/\text{d}$$

New WAS = 469 m³/d − 23 m³/d = 446 m³/d

$$\frac{446 \text{ m}^3}{1 \text{ d}} \times \frac{1000 \text{ L}}{1 \text{ m}^3} \times \frac{1 \text{ d}}{1440 \text{ min}} \times \frac{1 \text{ min}}{60 \text{ s}} = 5.16 \text{ L/s}$$

7.
$$\text{Sludge Age} = \frac{\text{Pounds of Bugs in Aeration}}{\text{Pounds/day of Bugs into Aeration}}$$

$$= \frac{\text{Pounds MLSS}}{\text{Pounds P.E. SS/day}}$$

$$\text{Sludge Age} = \frac{2150 \text{ mg/L} \times 0.54 \text{ MG} \times 8.34 \text{ lb/gal}}{128 \text{ mg/L} \times 3.14 \text{ MGD} \times 8.34 \text{ lb/gal}}$$

$$\text{Sludge Age} = \frac{9683 \text{ lb}}{3352 \text{ lb/d}}$$

$$= 2.89 \text{ d}$$

8.
$$\text{Sludge Age} = \frac{\text{Kilograms of Bugs in Aeration}}{\text{Kilograms/day of Bugs into Aeration}}$$

$$= \frac{\text{Kilograms MLSS}}{\text{Kilograms P.E. SS/day}}$$

$$\text{Sludge Age} = \frac{2150 \text{ mg/L} \times 2040 \text{ m}^3/1000}{128 \text{ mg/L} \times 11{,}890 \text{ m}^3/\text{d}/1000}$$

$$\text{Sludge Age} = \frac{4386 \text{ kg}}{1522 \text{ kg/d}}$$

$$= 2.88 \text{ d}$$

9.
$$\text{Solids Retention Time} = \frac{\text{Pounds of Bugs in Aeration}}{\text{Pounds/day of Bugs Leaving}}$$

$$= \frac{\text{Pounds MLSS}}{\text{Pounds/d WAS SS} + \text{Pounds/d S.E. SS}}$$

$$\text{SRT} = \frac{2180 \text{ mg/L} \times 0.48 \text{ MG} \times 8.34 \text{ lb/gal}}{(5340 \text{ mg/L} \times 0.095 \text{ MGD} \times 8.34 \text{ lb/gal}) + (17 \text{ mg/L} \times 2.92 \text{ MGD} \times 8.34 \text{ lb/gal})}$$

$$= \frac{8727 \text{ lb}}{4231 \text{ lb/d} + 414 \text{ lb/d}}$$

$$= \frac{8727 \text{ lb}}{4645 \text{ lb/d}}$$

$$= 1.88 \text{ d}$$

10.
$$\text{Solids Retention Time} = \frac{\text{Kilograms of Bugs in Aeration}}{\text{Kilograms/day of Bugs Leaving}}$$

$$= \frac{\text{Kilograms MLSS}}{\text{Kilograms/d WAS SS} + \text{Kilograms/d S.E. SS}}$$

$$SRT = \frac{2180 \text{ mg/L} \times 1820 \text{ m}^3/1000}{(5340 \text{ mg/L} \times 360 \text{ m}^3/\text{d}/1000) \div (17 \text{ mg/L} \times 11{,}050 \text{ m}^3/\text{d}/1000)}$$

$$= \frac{3968 \text{ kg}}{1922 \text{ kg/d} + 188 \text{ kg}}$$

$$= \frac{3968 \text{ kg}}{2110 \text{ kg/d}}$$

$$= 1.88 \text{ d}$$

11. Solids Retention Time

$$= \frac{\text{Pounds of Bugs in Aeration and Clarifier}}{\text{Pounds/day of Bugs Leaving}}$$

$$= \frac{\text{Pounds MLSS} + \text{Pounds SS in Clarifier}}{\text{Pounds/d WAS SS} + \text{Pounds/d S.E. SS}}$$

$$SRT = \frac{(2040 \text{ mg/L} \times 3.08 \text{ MG} \times 8.34 \text{ lb/gal}) + (1500 \text{ mg/L} \times 0.23 \text{ MG} \times 8.34 \text{ lb/gal})}{(4980 \text{ mg/L} \times 0.154 \text{ MGD} \times 8.34 \text{ lb/gal}) + (16 \text{ mg/L} \times 6.82 \text{ MGD} \times 8.34 \text{ lb/gal})}$$

$$= \frac{52{,}402 \text{ lb} + 2877 \text{ lb}}{6396 \text{ lb/d} + 910 \text{ lb/d}}$$

$$= \frac{55{,}279 \text{ lb}}{7306 \text{ lb/d}}$$

$$= 7.57 \text{ d}$$

12. Solids Retention Time

$$= \frac{\text{Kilograms of Bugs in Aeration and Clarifier}}{\text{Kilograms/day of Bugs Leaving}}$$

$$= \frac{\text{Kilograms MLSS} + \text{Kilograms SS in Clarifier}}{\text{Kilograms/d WAS SS} + \text{Kilograms/d'S.E. SS}}$$

$$\text{SRT} = \frac{\begin{array}{c}(2040\,\text{mg/L} \times 11{,}660\ \text{m}^3/\text{d}/1000) + \\ (1500\,\text{mg/L} \times 870\ \text{m}^3/1000)\end{array}}{\begin{array}{c}(4980\,\text{mg/L} \times 583\ \text{m}^3/\text{d}/1000) + \\ (16\,\text{mg/L} \times 25{,}820\ \text{m}^3/\text{d}/1000)\end{array}}$$

$$= \frac{23{,}786\,\text{kg} + 1305\,\text{kg}}{2903\,\text{kg/d} + 413\,\text{kg/d}}$$

$$= \frac{25{,}091\,\text{kg}}{3{,}316\,\text{kg/d}}$$

$$= 7.57\,\text{d}$$

---

13. Set up the calculation for the SRT and solve for the wasting rate.

$$\text{SRT} = \frac{\text{Bugs under Aeration}}{\text{Bugs/day Leaving}}$$

$$= \frac{\text{Pounds MLSS}}{\text{Pounds/d WAS SS} + \text{Pounds/d Eff SS}}$$

$$2.59\,\text{d} = \frac{1980\,\text{mg/L} \times 0.56\,\text{MG} \times 8.34\,\text{lb/gal}}{\begin{array}{c}(5120\,\text{mg/L} \times \text{WAS MGD} \times 8.34\,\text{lb/gal}) + \\ (15\,\text{mg/L} \times 3.44\,\text{MGD} \times 8.34\,\text{lb/gal})\end{array}}$$

$$2.59\,\text{d} = \frac{9247\,\text{lb}}{(5120\,\text{mg/L} \times \text{WAS MGD} \times 8.34\,\text{lb/gal}) + 430\,\text{lb/d}}$$

$$2.59\,\text{d} \times [(5120\,\text{mg/L} \times \text{WAS MGD} \times 8.34\,\text{lb/gal}) + 430\,\text{lb/d}]$$

$$= 9247\,\text{lb}$$

$$[2.59\,\text{d} \times (5120\,\text{mg/L} \times \text{WAS MGD} \times 8.34\,\text{lb/gal})]$$

$$+ [2.59\,\text{d} \times 430\,\text{lb/d}] = 9247\,\text{lb}$$

$$[2.59\,\text{d} \times (5120\,\text{mg/L} \times \text{WAS MGD} \times 8.34\,\text{lb/gal})]$$

$$+ 1114\,\text{lb} = 9247\,\text{lb}$$

$$[2.59\,\text{d} \times (5120\,\text{mg/L} \times \text{WAS MGD} \times 8.34\,\text{lb/gal})] = 8133\,\text{lb}$$

$$5120 \text{ mg/L} \times \text{WAS MGD} \times 8.34 \text{ lb/gal} = 3140 \text{ lb/d}$$

$$\text{WAS MGD} = \frac{3140 \text{ lb/d}}{5120 \text{ mg/L} \times 8.34 \text{ lb/gal}}$$

$$\text{WAS MGD} = 0.074 \text{ MGD}$$

$$\frac{0.074 \text{ million gal}}{1 \text{ d}} \times \frac{1 \text{ d}}{1440 \text{ min}} \times \frac{1,000,000 \text{ gal}}{1 \text{ million gal}}$$

$$= \frac{51.4 \text{ gal}}{1 \text{ min}} = 51.4 \text{ gpm}$$

14. Set up the calculation for the SRT and solve for the wasting rate.

$$\text{SRT} = \frac{\text{Bugs under Aeration}}{\text{Bugs/day Leaving}}$$

$$= \frac{\text{Kilograms MLSS}}{\text{Kilograms/d WAS SS} + \text{Kilograms/d Eff SS}}$$

$$2.59 \text{ d} = \frac{1980 \text{ mg/L} \times 2120 \text{ m}^3/1000}{(5120 \text{ mg/L} \times \text{WAS m}^3/\text{d}/1000) + (15 \text{ mg/L} \times 13,020 \text{ m}^3/\text{d}/1000)}$$

$$2.59 \text{ d} = \frac{4198 \text{ kg}}{(5120 \text{ mg/L} \times \text{WAS m}^3/\text{d}/1000) + 195 \text{ kg/d}}$$

$$2.59 \text{ d} \times [(5120 \text{ mg/L} \times \text{WAS m}^3/\text{d}/1000) + 195 \text{ kg/d}]$$

$$= 4198 \text{ kg}$$

$$[2.59 \times (5120 \text{ mg/L} \times \text{WAS m}^3/\text{d}/1000)] + [2.59 \text{ d} \times 195 \text{ kg/d}]$$

$$= 4198 \text{ kg}$$

$$[2.59 \text{ d} \times (5120 \text{ mg/L} \times \text{WAS m}^3/\text{d}/1000)] + 505 \text{ kg} = 4198 \text{ kg}$$

$$2.59 \text{ d} \times (5120 \text{ mg/L} \times \text{WAS m}^3/\text{d}/1000) = 3693 \text{ kg}$$

$$5120 \text{ mg/L} \times \text{WAS m}^3/\text{d}/1000 = 1426 \text{ kg/d}$$

$$\text{WAS m}^3/\text{d} = \frac{1426 \text{ kg/d} \times 1000}{5120 \text{ mg/L}}$$

$$\text{WAS m}^3/\text{d} = 278.5 \text{ m}^3/\text{d}$$

$$\frac{278.5 \text{ m}^3}{1 \text{ d}} \times \frac{1 \text{ d}}{1440 \text{ min}} \times \frac{1 \text{ min}}{60 \text{ s}} \times \frac{1000 \text{ L}}{1 \text{ m}^3} = 3.22 \text{ L/s}$$

15. Set up the calculation for the SRT and solve for the wasting rate.

Solids Retention Time

$$= \frac{\text{Pounds of Bugs in Aeration and Clarifier}}{\text{Pounds/day of Bugs Leaving}}$$

$$= \frac{\text{Pounds MLSS} + \text{Pounds SS in Clarifier}}{\text{Pounds/d WAS SS} + \text{Pounds/d S.E. SS}}$$

$$7.4 \text{ d} = \frac{\begin{array}{c}(2131 \text{ mg/L} \times 3.08 \text{ MG} \times 8.34 \text{ lb/gal}) + \\ (1710 \text{ mg/L} \times 0.23 \text{ MG} \times 8.34 \text{ lb/gal})\end{array}}{\begin{array}{c}(5060 \text{ mg/L} \times \text{WAS MGD} \times 8.34 \text{ lb/gal}) + \\ (14 \text{ mg/L} \times 6.78 \text{ MGD} \times 8.34 \text{ lb/gal})\end{array}}$$

$$7.4 \text{ d} = \frac{54,739 \text{ lb} + 3280 \text{ pounds}}{(5060 \text{ mg/L} \times \text{WAS MGD} \times 8.34 \text{ lb/gal}) + 792 \text{ lb/d}}$$

$$7.4 \text{ d} \times [(5060 \text{ mg/L} \times \text{WAS MGD} \times 8.34 \text{ lb/gal})] + 792 \text{ lb/d}]$$

$$= 58,019 \text{ lb}$$

$$[7.4 \text{ d} \times (5060 \text{ mg/L} \times \text{WAS MGD} \times 8.34 \text{ lb/gal})]$$

$$+ [7.4 \text{ d} \times 792 \text{ lb/d}] = 58,019 \text{ lb}$$

$$[7.4 \text{ d} \times (5060 \text{ mg/L} \times \text{WAS MGD} \times 8.34 \text{ lb/gal}) + 5,861 \text{ lb}$$

$$= 58,019 \text{ lb}$$

$$7.4 \text{ d} \times (5060 \text{ mg/L} \times \text{WAS MGD} \times 8.34 \text{ lb/gal}) = 52,158 \text{ lb}$$

$$5060 \text{ mg/L} \times \text{WAS MGD} \times 8.34 \text{ lb/gal} = 7048 \text{ lb/d}$$

$$\text{WAS MGD} = \frac{7048 \text{ lb/d}}{5060 \text{ mg/L} \times 8.34 \text{ lb/gal}}$$

$$\text{WAS MGD} = 0.167 \text{ MGD}$$

$$\frac{0.167 \text{ million gal}}{1 \text{ d}} \times \frac{1 \text{ d}}{1440 \text{ min}} \times \frac{1,000,000 \text{ gal}}{1 \text{ million gal}}$$

$$= \frac{116 \text{ gal}}{1 \text{ min}} = 116 \text{ gpm}$$

16. Set up the calculation for the SRT and solve for the wasting rate.

   Solids Retention Time

$$= \frac{\text{Kilograms of Bugs in Aeration and Clarifier}}{\text{Kilograms/day of Bugs Leaving}}$$

$$= \frac{\text{Kilograms MLSS} + \text{Kilograms SS in Clarifier}}{\text{Kilograms/d WAS SS} + \text{Kilograms/d S.E. SS}}$$

$$7.4 \text{ d} = \frac{\begin{array}{c}(2131 \text{ mg/L} \times 11{,}660 \text{ m}^3/1000) + \\ (1710 \text{ mg/L} \times 870 \text{ m}^3/1000)\end{array}}{\begin{array}{c}(5060 \text{ mg/L} \times \text{WAS m}^3/\text{d}/1000) + \\ (14 \text{ mg/L} \times 25{,}660 \text{ m}^3/\text{d}/1000)\end{array}}$$

$$7.4 \text{ d} = \frac{24{,}847 \text{ kg} + 1488 \text{ kg}}{(5060 \text{ mg/L} \times \text{WAS m}^3/\text{d}/1000) + 359 \text{ kg/day}}$$

$$7.4 \text{ d} \times [(5060 \text{ mg/L} \times \text{WAS m}^3/\text{d}/1000) + 359 \text{ kg/day}]$$

$$= 26{,}335 \text{ kg}$$

$$[7.4 \text{ d} \times (5060 \text{ mg/L} \times \text{WAS m}^3/\text{d}/1000)] + [7.4 \text{ d} \times 359 \text{ kg/d}]$$

$$= 26{,}335 \text{ kg}$$

$$[7.4 \text{ d} \times (5060 \text{ mg/L} \times \text{WAS m}^3/\text{d}/1000)] + 2657 \text{ kg} = 26{,}335 \text{ kg}$$

$$7.4 \text{ d} \times (5060 \text{ mg/L} \times \text{WAS m}^3/\text{d}/1000) = 23{,}678 \text{ kg}$$

$$5060 \text{ mg/L} \times \text{WAS m}^3/\text{d}/1000 = 3200 \text{ kg/d}$$

$$\text{WAS m}^3/\text{d} = \frac{3200 \text{ kg/d} \times 1000}{5060 \text{ mg/L}}$$

$$\text{WAS m}^3/\text{d} = 632.4 \text{ m}^3/\text{d}$$

$$\frac{632.4 \text{ m}^3}{1 \text{ d}} \times \frac{1 \text{ d}}{1440 \text{ min}} \times \frac{1 \text{ min}}{60 \text{ s}} \times \frac{1000 \text{ L}}{1 \text{ m}^3} = 7.32 \text{ L/s}$$

17.
$$\frac{F}{M} = \frac{\text{Pounds of Food}}{\text{Pounds of Bugs}} = \frac{\text{Pounds BOD (in)}}{\text{Pounds MLSS}}$$

Pounds BOD (in)

$$= 172 \text{ mg/L} \times 2.56 \text{ MGD} \times 8.34 \text{ lb/gal} = 3672 \text{ lb/d}$$

Pounds MLSS

$$= 2230 \text{ mg/L} \times 0.48 \text{ MG} \times 8.34 \text{ lb/gal} = 8927 \text{ lb}$$

$$\frac{F}{M} = \frac{3672 \text{ lb/d}}{8927 \text{ lb}} = 0.41$$

18.
$$\frac{F}{M} = \frac{\text{Kilograms of Food}}{\text{Kilograms of Bugs}} = \frac{\text{Kilograms BOD (in)}}{\text{Kilograms MLSS}}$$

Kilograms BOD (in)

$$= 172 \text{ mg/L} \times 9690 \text{ m}^3\text{/d/1000} = 1667 \text{ kg/d}$$

Kilograms MLSS

$$= 2230 \text{ mg/L} \times 1820 \text{ m}^3\text{/1000} = 4059 \text{ kg}$$

$$\frac{F}{M} = \frac{1667 \text{ kg/d}}{4059 \text{ kg}} = 0.41$$

19. Concentration of Suspended Solids in mg/L.

Sample Volume = 25 mL

	After Drying
Weight of Sample + Crucible	23.4791 g
Weight of Crucible	−23.4256 g
	0.0535 g (Weight of Dried Solids)

The unit desired is mg/L. This is a mass in some type of gram unit on the top and a volume in some type of liter unit on the bottom. Put what you do have in the place where you want it and just convert the specific units to what you need.

Want: mg/L   Have: grams and milliliters

We need to put the type of grams we do have on the top and the type of liters we do have on the bottom.

$$\frac{0.0535\ g}{25\ mL} \times \frac{1000\ mL}{1\ L} \times \frac{1000\ mg}{1\ g} = \frac{2140\ mg}{1\ L} = 2140\ mg/L$$

Percentage of the Volatile Suspended Solids in the Suspended Solids
When dry solids are burned, the solids can be viewed as separating into the material that has burned off or "volatilized" and the remaining ash.

$$Dry\ Solids = Volatile\ Solids + Ash$$

$$Dry\ Solids - Ash = Volatile\ Solids$$

	After Burning
Weight of Sample + Crucible	23.4406 g
Weight of Crucible	−23.4256 g
	0.0150 g (Weight of Remaining Ash)

Dry Solids − Ash = Volatile Solids

0.0535 g − 0.0150 g = 0.0385 g (Weight of Volatile Solids)

The percentage of volatile solids is: $\dfrac{Volatile\ Solids}{Dry\ Solids} \times 100\ \%$

$$\frac{0.0385\ g}{0.0535\ g} \times 100\ \% = 72\ \%$$

Concentration of Volatile Suspended Solids in mg/L

The concentration can be calculated directly as was done for the suspended solids concentration.

$$\frac{0.0385\ g}{25\ ml} \times \frac{1000\ mL}{1\ L} \times \frac{1000\ mg}{1\ g} = \frac{1540\ mg}{1\ L} = 1540\ mg/L$$

It is more common to see the VSS concentration calculated from the product of the SS concentration and the % VSS.

$$\text{VSS mg/L} = \text{SS mg/L} \times \frac{\% \text{ VSS}}{100 \%}$$

$$= 2140 \text{ mg/L} \times \frac{72 \%}{100 \%} = 1540 \text{ mg/L}$$

20. Calculate the BOD/COD ratio for each date listed and average the values.

$174/106 = 0.61$
$157/80 = 0.51$
$211/99 = 0.47$
$186/110 = 0.59$
$163/88 = 0.54$

$2.72/5 = 0.54$

$$\text{COD} \times \frac{\text{Average BOD}}{\text{COD}} = \text{Estimated BOD}$$

$$180 \text{ mg/L} \times 0.54 = 97 \text{ mg/L (Estimated BOD)}$$

21. Estimated BOD $= 0.58 \times 184 \text{ mg/L}$

$$= 107 \text{ mg/L}$$

$$\frac{F}{M} = \frac{107 \text{ mg/L} \times 2.64 \text{ MGD} \times 8.34 \text{ lb/gal}}{1940 \text{ mg/L} \times 0.48 \text{ MG} \times 8.34 \text{ lb/gal}}$$

$$\frac{F}{M} = \frac{2356 \text{ pounds/d}}{7766 \text{ pounds}}$$

$$\frac{F}{M} = 0.30$$

22.   Estimated BOD $= 0.58 \times 184 \text{ mg/L}$

$$= 107 \text{ mg/L}$$

$$\frac{F}{M} = \frac{107 \text{ mg/L} \times 10,000 \text{ m}^3/\text{d}/1000}{1940 \text{ mg/L} \times 1820 \text{ m}^3/1000}$$

$$\frac{F}{M} = \frac{1070 \text{ kg/d}}{3531 \text{ kg}}$$

$$\frac{F}{M} = 0.30$$

---

23. The way to approach this problem is to use the target F/M to determine a target MLSS concentration. Then use this MLSS concentration to calculate the change in the wasting rate.

$$\frac{F}{M} = \frac{\text{Pounds BOD (in)}}{\text{Pounds MLSS}}$$

$$0.42 = \frac{158 \text{ mg/L} \times 3.18 \text{ MGD} \times 8.34 \text{ lb/gal}}{\text{Target MLSS mg/L} \times 0.58 \text{ MG} \times 8.34 \text{ lb/gal}}$$

$$0.42 = \frac{4190 \text{ lb/d}}{\text{Target MLSS mg/L} \times 0.58 \text{ MG} \times 8.34 \text{ lb/gal}}$$

$$\text{Target MLSS mg/L} = \frac{4190 \text{ lb/d}}{0.42 \times 0.58 \text{ MG} \times 8.34 \text{ lb/gal}}$$

$$= 2062 \text{ mg/L}$$

We have: MLSS = 2090 mg/L
We want: MLSS = 2062 mg/L

$$28 \text{ mg/L (in excess)}$$

$$28 \text{ mg/L} \times 0.58 \text{ MG} \times 8.34 \text{ lb/gal} = 135 \text{ lb}$$

$$135 \text{ lb/d} = 5140 \text{ mg/L} \times \text{Additional WAS (MGD)} \times 8.34 \text{ lb/gal}$$

$$\frac{135}{5140 \times 8.34} = 0.003 \text{ MGD (Additional WAS)}$$

New WAS = 0.122 MGD + 0.003 MGD = 0.125 MGD

---

24. The way to approach this problem is to use the target F/M to determine a target MLSS concentration. Then use this MLSS concentration to calculate the change the wasting rate.

$$\frac{F}{M} = \frac{\text{Kilograms BOD (in)}}{\text{Kilograms MLSS}}$$

$$0.42 = \frac{158 \text{ mg/L} \times 12{,}040 \text{ m}^3/\text{d}/1000}{\text{Target MLSS mg/L} \times 2200 \text{ m}^3/1000}$$

$$0.42 = \frac{1902 \text{ kg/d}}{\text{Target MLSS mg/L} \times 2200 \text{ m}^3/1000}$$

$$\text{Target MLSS mg/L} = \frac{1902 \text{ kg/d}}{0.42 \times 2200 \text{ m}^3/1000}$$

$$= 2058 \text{ mg/L}$$

We have: MLSS = 2090 mg/L
We want: MLSS = 2058 mg/L

$$32 \text{ mg/L (in excess)}$$

$$32 \text{ mg/L} \times 2200 \text{ m}^3/1000 = 70 \text{ kg}$$

$$70 \text{ kg/d} = 5140 \text{ mg/L} \times \text{Additional WAS (m}^3\text{/d)}/1000$$

$$\frac{70 \times 1000}{5140} = 14 \text{ m}^3/\text{d (Additional WAS)}$$

$$\text{New WAS} = 462 \text{ m}^3/\text{d} + 14 \text{ m}^3/\text{d} = 476 \text{ m}^3/\text{d}$$

25. a. $\text{SRT} = \dfrac{(2560 \times 1.38 \times 8.34) + (1900 \times 0.117 \times 8.34)}{(5960 \times 0.075 \times 8.34) + (20 \times 2.9 \times 8.34)}$

$$= 7.4 \text{ d}$$

   b. The foaming is probably due to nocardia bacteria. If the SRT had been between nine and ten days, an increase in the wasting rate to reduce the SRT below nine days might have been considered. This type of nocardia may be the type that thrives at any SRT over two days. A change in wasting rate won't help the nocardia problem but might hurt the process. The foam may need to be removed by vacuum truck. Plant grease removal must be improved. If high influent grease concentrations are the problem, check to see if they are coming from one or two specific industries and enforce pretreatment compliance.

26. a. $\text{SRT} = \dfrac{(2560 \times 5220/1000) + (1900 \times 442/1000)}{(5960 \times 284 /1000) + (20 \times 10{,}980/1000)}$

$$= 7.4 \text{ d}$$

b. The foaming is probably due to nocardia bacteria. If the SRT had been between nine and ten days, an increase in the wasting rate to reduce the SRT below nine days might have been considered. This type of nocardia may be the type that thrives at any SRT over two days. A change in wasting rate won't help the nocardia problem but might hurt the process. The foam may need to be removed by vacuum truck. Plant grease removal must be improved. If high influent grease concentrations are the problem, check to see if they are coming from one or two specific industries and enforce pretreatment compliance.

ANSWERS TO THE QUESTIONS: NITRIFICATION AND DENTRIFICATION

1. Organic material can be thought of as long chains of carbon atoms.

2. An ion is an atom or a group of atoms held together and carrying a magnetic charge.

3. An ammonium ion has a single positive charge on it.

4. An ammonium ion is a nitrogen atom surrounded by four hydrogen atoms.

5. The form of nitrogen often highest in concentration in the influent of most municipal wastewater treatment plants is organic nitrogen.

6. The two different forms of ammonia are the ammonium ion (NH_4^+) and ammonia gas (NH_3).

7. These forms change back and forth depending on the pH of the water.

8. At a pH of 7.0, 99.9% of the ammonia is the ammonium ion (NH_4^+).

9. Nitrosomonas is the nitrogen bug that eats ammonia.

10. Nitrobacter is the nitrogen bug that eats nitrite.

11. Nitrification is the process that changes ammonia to nitrate.

12. Organic nitrogen can convert to ammonia. So the combination of organic nitrogen and ammonia could be called total potential ammonia. This test is done in the lab and is given the name Total Kjeldahl Nitrogen or TKN.

13. The term used to describe a tank condition with no dissolved oxygen present but with some nitrate present is anoxic.

14. Two examples of chemicals containing chemically combined oxygen are nitrite and nitrate.

15. Denitrification refers to the process in which certain carbon eaters will break down nitrate for its oxygen if no dissolved oxygen is in the water. The nitrogen from the nitrate forms nitrogen gas.

16. Denitrification is not desirable in a clarifier because the nitrogen bubbles will form inside floc particles and cause them to float and go over the weirs of the clarifier.

17. The nitrifiers are older than the carbon eating small bugs.

18. The carbon eating bugs come from domestic sewage.

19. The nitrifiers come only from the soil.

20. Alkalinity is a number of chemicals that collectively absorb acid.

21. Alkalinity is expressed in terms of equivalent mg/L of calcium carbonate.

22. Over seven pounds of alkalinity are needed for every pound of ammonia changed to nitrate.

23. The effluent looks cloudy if there are too many nitrifiers and not enough of the carbon eaters in the mixed liquor.

24. It is best to operate above a BOD:TKN ratio of 2.

25. The BOD:TKN ratio can be raised by bypassing primary effluent around first stage secondary treatment.

ANSWERS TO THE QUESTIONS: BIOLOGICAL NUTRIENT REMOVAL

1. Aerobic bugs require oxygen to live.

2. Anaerobic bugs live strictly without oxygen.

3. Facultative bugs can live with or without oxygen.

4. The facultative bugs remove the most amount of phosphorus.

5. The operator should provide the facultative bugs with food in the form of extra BOD under anaerobic conditions to give them a bigger advantage over the aerobic bugs.

6. The two categories of biological phosphorus removal processes are main stream and side stream processes.

7. The change in plant operation that some newer facilities are doing to help with biological phosphorus removal is not using primary settling tanks.

8. Nitrate in the return sludge can cause problems in the anaerobic zone by carrying chemically combined oxygen. By doing this, the zone will not be anaerobic and the facultative bugs will have no advantage.

9. When a plant is using a main stream biolgical phosphorus removal process the concern about the solids handling recycles is due to the fact

that the sludge has three times as much phosphorus. As the sludge is treated, this extra phosphorus will be put back into the water and returned to the wastewater flow unless it is chemically treated.

10. The facultative bugs require more energy to eat in anaerobic conditions.

11. A detention time of twelve hours is typically used in a stripper.

12. A return flow is settled sludge being returned to the head of the entire process. A recirculated flow involves mixed liquor being taken from one tank and pumped to another tank upstream.

SOLUTION TO CHALLENGE PROBLEM: BIOLOGICAL NUTRIENT REMOVAL

a. There are three flows going though the first anoxic zone. These are the wastewater, the return sludge and the recirculated mixed liquor.

5.7 MGD	21,580 m³/d
2.8 MGD	10,600 m³/d
+ 2.2 MGD	+ 8330 m³/d
10.7 MGD	40,510 m³/d

$$\frac{2.0 \text{ MG}}{10.7 \text{ MGD}} \times \frac{24 \text{ hr}}{1 \text{ d}} = 4.5 \text{ hrs} \qquad \frac{7,571 \text{ m}^3}{40,510 \text{ m}^3/\text{d}} \times \frac{24 \text{ hr}}{1 \text{ d}} = 4.5 \text{ hr}$$

b. Looking at the effluent limits, both the total phosphorus and the ammonia limits are being met. The limit for total nitrogen is not. How do we know this?

Total Nitrogen = Nitrite + Nitrate + (Organic Nitrogen + Ammonia)

= Nitrite + Nitrate + TKN

TN = 0.09 mg/L + 7.42 mg/L + 2.46 mg/L

= 9.97 mg/L

The effluent limit on total nitrogen is 5.0 mg/L. Since the ammonia limit is being met we know that the process is nitrifying. However, since there isn't enough nitrate being converted to nitrogen gas, we know that the process is not denitrifying as well as is needed.

Additional denitrification will occur by increasing the recirculation of the aerated mixed liquor to the first anoxic tank. The minimum detention time in this tank is three hours. Now work the detention time calculation for this tank to find the maximum recirculation rate that could be used and still be above the required minimum detention time.

$$\frac{20\ MG\ \times\ 24\ hr/d}{5.7\ MGD\ +\ 2.8\ MGD\ +\ R\ MGD} = 3.0\ hrs$$

$$\frac{2.0\ MG\ \times\ 24\ hr/d}{3.0\ hrs} = 8.5\ MGD\ +\ R\ MGD$$

$$\frac{2.0\ MG\ \times\ 24\ hr/d}{3.0\ hrs} - 8.5\ MGD = R\ MGD$$

$$16.0\ MGD\ -\ 8.5\ MGD = R\ MGD$$

$$7.5\ MGD = R\ MGD$$

$$\frac{7571\ m^3\ \times\ 24\ hrs/d}{21,580\ m^3/d\ +\ 10,600\ m^3/+d\ +\ R\ m^3/d} = 3.0\ hrs$$

$$\frac{7571\ m^3\ \times\ 24\ hrs/d}{3.0\ hrs} = 32,180\ m^3/d\ +\ R\ m^3/d$$

$$\frac{7571\ m^3\ \times\ 24\ hrs/d}{3.0\ hrs} - 32,180\ m^3/d = R\ m^3/d$$

$$60,568\ m^3/d\ -\ 32,180\ m^3/d = R\ m^3/d$$

$$28,388\ m^3/d = R\ m^3/d$$

Recirculation of aeration mixed liquor can be increased up to 7.5 MGD or 28,388 m^3/d to still maintain the three hour minimum detention time. Increase the recirculation rate gradually. Monitor the concentration of total nitrogen in the effluent for permit compliance. The largest concern with increasing the recirculation rate is the amount of dissolved oxygen being carried into the anoxic zone with it. For this reason, only increase recirculation enough to meet the effluent permit.

ANSWERS TO THE QUESTIONS: WASTE TREATMENT PONDS

1. The three categories of lagoons are aerobic, facultative, and anaerobic.

2. The most common type of pond is the facultative pond.

3. The two most common sources of oxygen for lagoons are algae and surface aerators.

4. The problem with effluent quality that the use of algae causes is increased concentration of suspended solids from algae pieces.

5. The solutions that could be considered to solve this algae problem include installation of an effluent baffle, dosage of alum, or installation of effluent sand filters.

6. The depth of a facultative lagoon is usually between 3 and 6 ft.

7. A lack of wave action on the lagoon surface might indicate an oily surface or anaerobic conditions.

8. When carbon dioxide is added to water the pH will go down because of the formation of carbonic acid.

9. During a 24-hour cycle in the summer the algae will produce oxygen and reduce carbon dioxide during the day and produce carbon dioxide

and reduce oxygen at night. This will cause the DO and pH to rise during the day and go down at night.

10. A dark green lagoon color indicates higher DO concentration and pH.

11. A dull green to yellow color indicates the development of blue-green algae.

12. Muskrats can be discouraged from burrowing around a lagoon by raising and lowering the water level by a few inches over a couple of weeks.

13. The chemical that can be added to a lagoon to increase available oxygen is sodium nitrate.

14. The purpose of an anaerobic lagoon is just to stabilize solids.

15. The odors from an anaerobic lagoon can be reduced when the unit has just recently been started up by placing sheets of thin styrofoam across the tank surface until a scum blanket develops.

SOLUTIONS TO THE PROBLEMS: WASTE TREATMENT PONDS

1.
$$705 \text{ ft} \times 450 \text{ ft} \times 5.8 \text{ ft} \times \frac{7.48 \text{ gal}}{1 \text{ ft}^3} = 13,760,000 \text{ gal}$$

$$13,760,000 \text{ gal} \times \frac{1 \text{ d}}{290,000 \text{ gal}} = 47.5 \text{ d}$$

2.
$$215 \text{ m} \times 135 \text{ m} \times 1.8 \text{ m} \times \frac{1 \text{ d}}{1100 \text{ m}^3} = 47.5 \text{ d}$$

3.
$$695 \text{ ft} \times 425 \text{ ft} \times 49 \text{ in.} \times \frac{1 \text{ ft}}{12 \text{ in.}} = 1,206,000 \text{ ft}^3$$

$$\frac{0.45 \text{ ac-ft}}{1 \text{ d}} \times \frac{43,560 \text{ ft}^3}{1 \text{ ac-ft}} = \frac{19,602 \text{ ft}^3}{1 \text{ d}}$$

$$1,206,000 \text{ ft}^3 \times \frac{1 \text{ d}}{19,602 \text{ ft}^3} = 61.5 \text{ d}$$

4.
$$\frac{49 \text{ in.}}{61.5 \text{ d}} = 0.80 \text{ in.} /\text{d}$$

5.
$$\frac{1.8 \text{ m} \times \dfrac{100 \text{ cm}}{1 \text{ m}}}{47.5 \text{ d}} = \frac{3.79 \text{ cm}}{1 \text{ d}}$$

6.
$$0.14 \text{ MGD} \times 162 \text{ mg/L} \times 8.34 \text{ lb/gal} = 189.2 \text{ lb/d}$$

$$692 \text{ ft} \times 387 \text{ ft} \times \frac{1 \text{ ac}}{43,560 \text{ ft}^2} = 6.15 \text{ ac}$$

$$\frac{189.2 \text{ lb/d}}{61.5 \text{ ac}} = 30.8 \text{ lb/d/ac}$$

7.
$$\frac{559 \text{ m}^3/\text{d} \times 162 \text{ mg/L}}{1000} = 90.56 \text{ kg/d}$$

$$206 \text{ m} \times 118 \text{ m} \times \frac{1 \text{ ha}}{10,000 \text{ m}^2} = 2.43 \text{ ha}$$

$$\frac{90.56 \text{ kg/d}}{2.43 \text{ ha}} = 37.3 \text{ kg/d/ha}$$

8.
$$742 \text{ ft} \times 426 \text{ ft} \times \frac{1 \text{ ac}}{43,560 \text{ ft}^2} = 7.26 \text{ ac}$$

$$\frac{0.62 \text{ ac-ft/d}}{7.26 \text{ ac}} \times \frac{12 \text{ in.}}{1 \text{ ft}} = 1.02 \text{ in./d}$$

9.
$$227 \text{ m} \times 124 \text{ m} = 28,148 \text{ m}^2$$

$$\frac{792 \text{ m}^3/\text{d}}{28,148 \text{ m}^2} \times \frac{100 \text{ cm}}{1 \text{ m}} = 2.81 \text{ cm/d}$$

10. a.
$$670 \text{ ft} \times 450 \text{ ft} \times \frac{1 \text{ ac}}{43,560 \text{ ft}^2} = 6.92 \text{ ac}$$

$$\frac{100 \text{ lb}}{1 \text{ ac}} \times 6.92 \text{ ac} = 692 \text{ lb}$$

$$\frac{50 \text{ lb}}{1 \text{ ac}} \times 6.92 \text{ ac} \times 6 \text{ d} = 2076 \text{ lb}$$

$$692 \text{ lb} + 2076 \text{ lb} = 2768 \text{ lb}$$

b. Supplement the aeration with surface aerators or by use of a motor boat. Recirculate a portion of the pond effluent. Eliminate septic or high strength pond influent.

---

11. a.

$$204 \text{ m} \times 137 \text{ m} \times \frac{\text{ha}}{10,000 \text{ m}^2} = 2.79 \text{ ha}$$

$$\frac{112 \text{ kg}}{1 \text{ ha}} \times 2.79 \text{ ha} = 312.5 \text{ kg}$$

$$\frac{56 \text{ kg}}{1 \text{ ha}} \times 2.79 \text{ ha} \times 6 \text{ d} = 937.5 \text{ kg}$$

$$312.5 \text{ kg} + 937.5 \text{ kg} = 1250 \text{ kg}$$

b. Supplement the aeration with surface aerators or by use of a motor boat. Recirculate a portion of the pond effluent. Eliminate septic or high strength pond influent.

ANSWERS TO THE QUESTIONS: PHYSICAL AND CHEMICAL TREATMENT

1. Four chemicals that can be added to wastewater to remove phosphorus are lime, alum, ferric chloride, and pickle liquor.

2. The unit used to mix lime with water is called a slaker.

3. When alum is added to the treatment process the sludge becomes more difficult to dewater.

4. Before pickle liquor was green it was either hydrochloric or sulfuric acid.

5. The two categories of filters are single and multiple media.

6. Multiple media filters are better for wastewater treatment.

7. The media size distribution in a multiple media filter ranges from the largest at the top of the bed to the smallest at the bottom.

8. The solids in a single media filter collect at the surface.

9. The solids in a multiple media filter are distributed throughout the filter.

10. The flow passes downward through the filter during normal operation.

11. The two parameters used to monitor filter operation are effluent turbidity and the filtered wastewater pressure differential.

12. Polymer should be added directly at the influent to the filter.

13. If the polymer dosage is too high the flocs will be too large to penetrate the bed. They will collect on the top of the filter and quickly blind it.

14. If the polymer dosage is too low the flocs will be too small and pass through the bed. The effluent turbidity will rise quickly after start-up.

15. As water temperature rises the required backwash flowrate goes up.

16. As the temperature goes up the amount of polymer needed goes down.

17. The percentage of backwash that should not be exceeded is 5%.

SOLUTIONS TO THE PROBLEMS: PHYSICAL AND CHEMICAL TREATMENT

1.
$$\text{Volume backwash water} = \frac{8420 \text{ gal}}{1 \text{ min}} \times 9 \text{ min} \times 6 \text{ filters}$$

$$= 454,680 \text{ gal}$$

$$\text{Water filtered since last BW} = (1440 \text{ min} - 9 \text{ min})$$

$$\times \frac{2730 \text{ gal}}{1 \text{ min}} \times 6 \text{ filters}$$

$$= 23,440,000 \text{ gal}$$

$$\text{Percent Backwash} = \frac{454,680 \text{ gal}}{23,440,000 \text{ gal}} \times 100\% = 1.9\%$$

2.
$$\text{Volume backwash water} = \frac{530 \text{ L}}{1 \text{ s}} \times \frac{60 \text{ sec}}{1 \text{ min}} \times 9 \text{ min} \times 6 \text{ filters}$$

$$= 1,717,200 \text{ L}$$

$$\text{Water filtered since last BW} = (1440 \text{ min} - 9 \text{ min}) \times \frac{172 \text{ L}}{1 \text{ s}}$$

$$\times \frac{60 \text{ s}}{\text{min}} \times 6 \text{ filters}$$

$$= 88,608,000 \text{ L}$$

$$\text{Percent Backwash} = \frac{1,717,200 \text{ L}}{88,608,000 \text{ L}} \times 100\% = 1.9\%$$

3.
$$\frac{14,200,000 \text{ gal}}{67 \text{ hr}} \times \frac{1 \text{ hr}}{60 \text{ min}} = 3532 \text{ gal/min}$$

$$34 \text{ ft} \times 22 \text{ ft} = 748 \text{ ft}^2$$

$$\frac{3532 \text{ gpm}}{748 \text{ ft}^2} = 4.72 \text{ gpm/ft}^2$$

4.
$$\frac{53,750 \text{ m}^3}{67 \text{ hr}} \times \frac{1 \text{ hr}}{60 \text{ min}} = 13.4 \text{ m}^3/\text{min}$$

$$10.4 \text{ m} \times 6.7 \text{ m} = 69.7 \text{ m}^2$$

$$\frac{13.4 \text{ m}^3/\text{min}}{69.7 \text{ m}^2} = 0.19 \text{ m}^3/\text{min}/\text{m}^2$$

5. a.
$$15 \text{ ft} \times 18 \text{ ft} \times \frac{24.2 \text{ in}}{1 \text{ min}} \times \frac{1 \text{ ft}}{12 \text{ in.}} \times \frac{7.48 \text{ gal}}{1 \text{ ft}^3} = \frac{4073 \text{ gal}}{\text{min}}$$

$$4.6 \text{ m} \times 5.5 \text{ m} \times \frac{61.5 \text{ cm}}{1 \text{ min}} \times \frac{1 \text{ m}}{100 \text{ cm}} \times \frac{1000 \text{ L}}{1 \text{ m}^3} \times \frac{1 \text{ min}}{60 \text{ sec}}$$

$$= 259.3 \text{ L/s}$$

b. Since February, the temperature of the water has increased. For this reason, the water with a lower viscosity does not expand the bed as much as it needs to be.

Media Depth = Sand + Coal = 62 in.	158 cm
− Sand Depth = 18 in.	46 cm
Coal Depth = 44 in.	112 cm

25% of coal depth × 0.25 = 11 in.	× 0.25 = 28 cm
− present expansion − 8 in.	− 20
3 in.	8 cm

The bed is only expanding by 8 in. or 20 cm during backwashes. The bed needs to expand by 11 in. or 28 cm in order to get clean. Increase the backwash flowrate until the needed bed expansion is obtained.

ANSWERS TO THE QUESTIONS: CHLORINATION AND DECHLORINATION

1. The bugs that are disease carriers are called pathogens.

2. The process of killing most of the bugs is called disinfection.

3. The amount of chlorine added is called the dosage.

4. The amount of chlorine that gets used up is called the demand.

5. The amount of chlorine that is left in the water after the demand has been satisfied is called the chlorine residual.

6. The form of chlorine used which could also be called a strong bleach is sodium hypochlorite.

7. The color of chlorine gas is greenish-yellow.

8. Chlorine vents should be placed near or at the floor because chlorine gas is heavier than air.

9. The chemical symbol for free available chlorine is HOCl.

10. Chlorine forms chloramines with ammonia.

11. We measure the amount of disease carrying bugs in the water by using the indicator organism — coliform.

12. Only enough chlorine should be added to reduce the coliform count of the treated effluent to within the permit limits.

13. Two chemicals used to dechlorinate are sulfur dioxide and sodium bisulfite.

14. An instrument that could be used to measure chlorine residual concentrations accurately to three decimal places is the spectrophotometer.

SOLUTIONS TO THE PROBLEMS: CHLORINATION AND DECHLORINATION

1.
$$\text{Dosage} = \text{Demand} + \text{Residual}$$

$$\text{Dosage} - \text{Residual} = \text{Demand}$$

$$8.9 \text{ mg/L} - 0.4 \text{ mg/L} = 8.5 \text{ mg/L}$$

2.
$$\text{Dosage} = \text{Demand} + \text{Residual}$$

$$= 7.8 \text{ mg/L} + 0.3 \text{ mg/L}$$

$$= 8.1 \text{ mg/L}$$

$$5.7 \text{ MGD} \times 8.1 \text{ mg/L} \times 8.34 \text{ lb/gal} = 385 \text{ lb/d}$$

3.
$$\text{Dosage} = \text{Demand} + \text{Residual}$$

$$= 7.8 \text{ mg/L} + 0.3 \text{ mg/L}$$

$$= 8.1 \text{ mg/L}$$

$$21.6 \text{ ML/d} \times 8.1 \text{ mg/L} = 175 \text{ kg/d}$$

SOLUTION TO CHALLENGE PROBLEM: CHLORINATION AND DECHLORINATION

The problem is due to the lack of ammonia in the nitrified effluent. Ideally, an effluent with a 1 to 2 mg/L ammonia concentration would be best. There would be enough ammonia for the chlorine to form the more efficient chloramines and most of the ammonia would be reduced in the process. So how do we get an effluent with a 1 to 2 mg/L ammonia concentration?

Maybe the activated sludge unit could be operated to convert less ammonia. This, unfortunately, does not work. Nitrification is like a light switch. It is either on or off. Any attempt to operate in between will not be stable.

Ammonia could be added to the effluent before chlorination. This is done at some treatment plants. However, it is a tradeoff of one chemical cost for another.

The best solution is to bypass about 10% of the trickling filter effluent around the activated sludge unit. The ammonia in this flow will provide the ammonia to form the chloramines. If no means to legitimately bypass exist, then consider allowing this 10% flow to go through an aeration tank without return sludge added. The results could easily be a 50% reduction in chlorine usage through a simple process change.

ANSWERS TO THE QUESTIONS: SLUDGE THICKENING

1. The four commonly used methods to thicken sludge are gravity thickening, dissolved air flotation, centrifugation, and gravity belt thickening.

2. A gravity thickener is better at thickening primary sludge.

3. The efficiency of a gravity thickener is improved when the influent to the unit is less concentrated.

4. The primary sludge can be made more fresh going into a gravity thickener by increasing the primary sludge pumping rate or by adding dilution water.

5. If the biological activity in settling secondary sludge becomes a problem, either chlorine or permanganate can be added to help.

6. Gravity thickeners should be skimmed regularly to reduce a potential nocardia foam accumulation.

7. Pickets are sometimes used in a gravity thickener to cut through the settling sludge, releasing gas bubbles which helps the sludge to settle better.

8. In a dissolved air flotation unit, the range of pressures used is from 40 to 70 psi.

9. An operator can expect to get a sludge concentration between 4 and 8% using a dissolved air flotation thickener.

10. It is better for the influent to a dissolved air flotation thickener to be less concentrated.

11. An operator can expect to get a sludge float blanket between 8 and 24 inches thick when operating a dissolved air flotation thickener.

12. The two most important factors that affect the operation of a centrifuge are the flowrate of the sludge going into the unit and the pounds or kilograms of solids in the influent.

13. The two different spinning parts in a solid bowl centrifuge are the bowl and the screw-conveyor.

14. The differential speed for a solid bowl centrifuge is the difference in rotational speeds between the bowl and the screw-conveyor.

15. Polymer is required in the operation of a gravity belt thickener.

16. If the size of the holes in the belt of a gravity belt thickener is too small, water will not drain. If the holes are too large, then sludge solids will pass through the holes.

SOLUTIONS TO THE PROBLEMS: SLUDGE THICKENING

1.
$$\frac{48,600 \text{ gal}}{d} \times \frac{8.34 \text{ lb}}{\text{gal}} \times \frac{0.9}{100} = \frac{3648 \text{ lb}}{d}$$

2.
$$\frac{184,000 \text{ L}}{d} \times \frac{1 \text{ kg}}{1 \text{ L}} \times \frac{0.9}{100} = \frac{1656 \text{ kg}}{d}$$

3.
$$\text{Area} = 0.785 \times (26 \text{ ft})^2$$

$$= 530.66 \text{ ft}^2$$

$$\text{Hydraulic Loading Rate} = \frac{162 \text{ gal}}{1 \text{ min}} \times \frac{1440 \text{ min}}{1 \text{ d}} \times \frac{1}{530.66 \text{ ft}^2}$$

$$= 440 \text{ gpd/ft}^2$$

4.
$$\text{Area} = 0.785 \times (7.9 \text{ m})^2$$

$$= 48.99 \text{ m}^2$$

$$\text{Hydraulic Loading Rate} = \frac{10.2 \text{ L}}{1 \text{ s}} \times \frac{60 \text{ s}}{1 \text{ min}} \times \frac{1440 \text{ min}}{1 \text{ d}}$$

$$\times \frac{1 \text{ m}^3}{1000 \text{ L}} \times \frac{1}{48.99 \text{ m}^2}$$

$$= 18.0 \text{ m}^3/\text{d}/\text{m}^2$$

5.
$$\frac{260 \text{ gal}}{1 \text{ min}} \times \frac{1440 \text{ min}}{1 \text{ d}} \times \frac{8.34 \text{ lb}}{1 \text{ gal}} \times \frac{1.2}{100} = 37{,}470 \text{ lb}/\text{d}$$

$$0.785 \times (42 \text{ ft})^2 = 1384.7 \text{ ft}^2$$

$$\frac{37{,}470 \text{ lb}/\text{d}}{1384.7 \text{ ft}^2} = 27.1 \text{ lb}/\text{d}/\text{ft}^2$$

6.
$$\frac{16.4 \text{ L}}{1 \text{ s}} \times \frac{60 \text{ s}}{1 \text{ min}} \times \frac{1440 \text{ min}}{1 \text{ d}} \times \frac{1 \text{ kg}}{1 \text{ L}} \times \frac{1.2}{100} = 17{,}004 \text{ kg}/\text{d}$$

$$0.785 \times (12.8 \text{ m})^2 = 128.6 \text{ m}^2$$

$$\frac{17{,}004 \text{ kg}/\text{d}}{128.6 \text{ m}^2} = 132.2 \text{ kg}/\text{d}/\text{m}^2$$

7.
$$\frac{680 \text{ gpm}}{0.785 \times (32 \text{ ft})^2} = 0.85 \text{ gpm}/\text{ft}^2$$

8.
$$\frac{42.5 \text{ L}/\text{s}}{0.785 \times (9.8 \text{ m})^2} = 0.56 \text{ L}/\text{s}/\text{m}^2$$

9.
$$\frac{125 \text{ gal}}{1 \text{ min}} \times \frac{60 \text{ min}}{1 \text{ hr}} \times \frac{8.34 \text{ lb}}{1 \text{ gal}} \times \frac{0.95}{100} = 594.2 \text{ lb}/\text{hr}$$

$$\text{Area} = 0.785 \times (32 \text{ ft})^2 = 803.84 \text{ ft}^2$$

$$\frac{594.2 \text{ lb}/\text{hr}}{803.84 \text{ ft}^2} = 0.74 \text{ lb}/\text{hr}/\text{ft}^2$$

10.
$$\frac{7.9 \text{ L}}{1 \text{ s}} \times \frac{60 \text{ sec}}{1 \text{ min}} \times \frac{60 \text{ min}}{1 \text{ hr}} \times \frac{1 \text{ kg}}{1 \text{ L}} \times \frac{0.95}{100} = 270.18 \text{ kg}/\text{hr}$$

$$\text{Area} = 0.785 \times (9.8 \text{ m})^2 = 75.4 \text{ m}^2$$

$$\frac{270.18 \text{ kg/hr}}{75.4 \text{ m}^2} = 3.58 \text{ kg/hr/m}^2$$

11. a. Do a balance of solids in and out of the unit.

Pounds of solids going into the thickener:

$$\frac{147 \text{ gal}}{1 \text{ min}} \times \frac{1440 \text{ min}}{1 \text{ d}} \times \frac{8.34 \text{ lb}}{1 \text{ gal}} \times \frac{2.6}{100} = 45,901 \text{ lb/d}$$

Pounds of solids leaving in the thickened sludge:

$$\frac{77 \text{ gal}}{1 \text{ min}} \times \frac{1440 \text{ min}}{1 \text{ d}} \times \frac{8.34 \text{ lb}}{1 \text{ gal}} \times \frac{5.2}{100} = 48,086 \text{ lb/d}$$

There are more pounds of solids leaving in the sludge than are entering the thickener. If nothing changes, the blanket level will decrease.

  b. If the thickened sludge starts to thin, decrease the pumping rate of the thickened sludge.

12. a. Do a balance of solids in and out of the unit.

Kilograms of solids going into the thickener:

$$\frac{9.2 \text{ L}}{1 \text{ s}} \times \frac{60 \text{ s}}{1 \text{ min}} \times \frac{1440 \text{ min}}{1 \text{ d}} \times \frac{1 \text{ kg}}{1 \text{ l}} \times \frac{2.6}{100} = 20,667 \text{ kg/d}$$

Kilograms of solids leaving in the thickened sludge:

$$\frac{4.8 \text{ L}}{1 \text{ s}} \times \frac{60 \text{ s}}{1 \text{ min}} \times \frac{1440 \text{ min}}{1 \text{ d}} \times \frac{1 \text{ kg}}{1 \text{ L}} \times \frac{5.2}{100} = 21,565 \text{ kg/d}$$

There are more solids leaving in the sludge than are entering the thickener. If nothing changes, the blanket level will decrease.

  b. If the thickened sludge starts to thin, decrease the pumping rate of the thickened sludge.

ANSWERS TO THE QUESTIONS: ANAEROBIC SLUDGE DIGESTION

1. There are two basic steps in the anaerobic digestion process.

2. The first step in the anaerobic digestion process is the conversion of organic materials to volatile acids.

3. The second step in the anaerobic digestion process is the conversion of the volatile acids to methane.

4. The bugs that do the first step in the anaerobic digestion process are sometimes called acid formers.

5. The bugs that do the second step in the anaerobic digestion process have been called methanogenic bugs or methane formers.

6. The pH is a measure of how much acid is in the water.

7. A value of 7.0 is a neutral pH.

8. If the anaerobic digester pH goes below 6.6, additional scum and odors may be generated.

9. If the anaerobic digester pH goes above 7.5, the bugs may slow down or even stop working.

10. Alkalinity is one measurement for a number of different chemicals that have the ability to absorb acid.

11. The range for mesophilic bugs is 80 to 100°F or 26 to 38°C.

12. The range for thermophilic bugs is 110 to 140°F or 43 to 60°C.

13. Most anaerobic digesters operate at 95°F or 35 °C.

14. Digester gas contains both carbon dioxide and methane.

15. The organic material in the raw sludge is measured by the total volatile solids concentration.

16. The explosive range of methane in air is approximately 5% to 20%.

17. The first indication that an anaerobic digester is having a problem is when the volatile acids concentration or the volatile acid to alkalinity ratio starts to increase.

18. The value of the volatile acid to alkalinity ratio for a properly operated anaerobic digester is generally less than 0.1.

19. The amount of carbon dioxide in the digester gas for a properly operated anaerobic digester is generally between 30% and 35%.

20. Four potential causes of an anaerobic digester operational problem are hydraulic and organic overloading, temperature fluctuation, and toxic chemicals.

21. The maximum percent volume that accumulated grit and scum should occupy in a digester is 5%.

22. The recommended schedule for feeding raw sludge to a digester is 5 to 10 min per hour.

23. The two chemicals that can be added to increase alkalinity are lime and sodium bicarbonate.

24. If the volatile acid to alkalinity ratio is at a value of 0.23 and rising the concentration of volatile acids is increasing which is definitely a problem.

SOLUTIONS TO THE PROBLEMS: ANAEROBIC SLUDGE DIGESTION

1.
$$\frac{3.4 \text{ gal}}{1 \text{ min}} \times \frac{1440 \text{ min}}{1 \text{ d}} \times \frac{8.34 \text{ lb}}{1 \text{ gal}} \times \frac{4.9}{100} \times \frac{69}{100}$$

$$= 1381 \text{ lb VS/day}$$

2. $$\frac{18,500 \text{ L}}{1 \text{ d}} \times \frac{1 \text{ kg}}{1 \text{ l}} \times \frac{4.9}{100} \times \frac{69}{100} = 625.5 \text{ kg VS/d}$$

3. First calculate the volume of sludge in the tank.

$$0.785 \times (52 \text{ ft})^2 \times 20 \text{ ft} = 42,453 \text{ ft}^3$$

Next calculate the organic loading in lb VS/day.

$$\frac{36,900 \text{ gal}}{1 \text{ d}} \times \frac{8.34 \text{ lb}}{1 \text{ gal}} \times \frac{5.4}{100} \times \frac{70}{100} = 11,633 \text{ lb VS/d}$$

Divide the two.

$$\frac{11,633 \text{ lb VS/d}}{42,453 \text{ ft}^3} = 0.27 \text{ lb VS/ft}^3/\text{d}$$

4. First calculate the volume of sludge in the tank.

$$0.785 \times (15.8 \text{ m})^2 \times 6.1 \text{ m} = 1195.4 \text{ m}^3$$

Next calculate the organic loading in kg VS/day.

$$\frac{139,600 \text{ L}}{1 \text{ d}} \times \frac{1 \text{ kg}}{1 \text{ L}} \times \frac{5.4}{100} \times \frac{70}{100} = 5276.9 \text{ kg VS/d}$$

Divide the two.

$$\frac{5276.9 \text{ kg VS/d}}{1195.4 \text{ m}^3} = 4.4 \text{ kg VS/m}^3/\text{d}$$

5. $$\frac{176 \text{ mg/L}}{2060 \text{ mg/L}} = 0.085$$

6. $$196,000 \text{ gal} = 0.196 \text{ MG}$$

$$0.196 \text{ MG} \times 8.34 \text{ lb/gal} \times 1840 \text{ mg/L} = 3008 \text{ lb}$$

7. $$745,000 \text{ L} = 0.745 \text{ ML}$$

$$0.745 \text{ ML} \times 1840 \text{ mg/L} = 1371 \text{ kg}$$

8.
$$\% \text{ VS Reduction} = \frac{\text{VS (in)} - \text{VS (out)}}{\text{VS (in)} - [\text{VS (in)} \times \text{VS (out)}]}$$

where $\text{VS} = \%\text{VS}/100$.

$$\% \text{ VS Reduction} = \frac{0.70 - 0.53}{0.70 - (0.70 \times 0.53)} \times 100\% = 51.7\%$$

9.
$$\% \text{ VS Reduction} = \frac{0.68 - 0.54}{0.68 - (0.68 \times 0.54)} \times 100\% = 44.8\%$$

10.
$$5{,}500 \text{ lb VS} \times \frac{100}{9.7} \times \frac{100}{68} \times \frac{1 \text{ gal}}{8.7 \text{ lb}} = 9{,}584 \text{ gal}$$

11.
$$2{,}500 \text{ kg VS} \times \frac{100}{9.7} \times \frac{100}{68} \times \frac{1 \text{ L}}{1.04 \text{ kg}} = 36{,}440 \text{ L}$$

12. First calculate the % VS reduction

$$\frac{0.68 - 0.52}{0.68 - (0.68 \times 0.52)} \times 100\% = 49.0\%$$

Next calculate the pounds of VS in the raw sludge.

$$\frac{84{,}000 \text{ gal}}{1 \text{ d}} \times \frac{5.4}{100} \times \times \frac{68}{100} \times \frac{8.34 \text{ lb}}{1 \text{ gal}} = 25{,}725 \text{ lb/d}$$

Then calculate the pounds of VS destroyed.

$$\frac{25{,}725 \text{ lb}}{1 \text{ d}} \times \frac{49}{100} = \frac{12{,}605 \text{ lb destroyed}}{1 \text{ d}}$$

Finally calculate the gas production.

$$\frac{12{,}605 \text{ lb}}{1 \text{ d}} \times \frac{12 \text{ ft}^3}{1 \text{ lb destroyed}} = 151{,}263 \text{ ft}^3/\text{d}$$

13. First calculate the % VS reduction

$$\frac{0.68 - 0.52}{0.68 - (0.68 \times 0.52)} \times 100\% = 49.0\%$$

Next calculate the pounds of VS in the raw sludge.

$$\frac{318,000 \text{ L}}{1 \text{ d}} \times \frac{5.4}{100} \times \frac{68}{100} \times \frac{1 \text{ kg}}{1 \text{ L}} = 11,677 \text{ kg/d}$$

Then calculate the pounds of VS destroyed.

$$\frac{11,677 \text{ kg}}{1 \text{ d}} \times \frac{49}{100} = \frac{5722 \text{ kg destroyed}}{1 \text{ d}}$$

Finally calculate the gas production

$$\frac{5722 \text{ kg}}{1 \text{ d}} \times \frac{0.75 \text{ m}^3}{1 \text{ kg destroyed}} = 4291.5 \text{ m}^3/\text{d}$$

14. a.

Week	Day	Volatile Acids (mg/L)	Alkalinity (mg/L)	VA/Alkalinity
#1	Mon	142	1990	0.071
	Wed	145	1830	0.079
	Fri	154	1690	0.091
#2	Mon	150	1810	0.083
	Wed	151	1800	0.084
	Fri	149	1910	0.078
#3	Mon	150	1850	0.081
	Wed	155	1560	0.099
	Fri	160	1430	0.112
#4	Mon	167	1246	0.134
	Wed.	174	1140	0.153
	Fri	189	1070	0.177

b. The VA/Alkalinity ratio is starting to increase. This may be the start of a problem.

Check the digester temperature for stability. The raw sludge feed may be lowering the digester temperature by too much. Try feeding smaller amounts more frequently or preheating the raw sludge if possible.

Decreasing the daily raw sludge feed as well as digested sludge withdrawal will give the solids more detention time to stabilize. Try correcting any operational problems early and naturally before considering larger changes or chemical addition.

ANSWERS TO THE QUESTIONS: SLUDGE DEWATERING

1. The three general ways to condition sludge are the addition of inorganic chemicals, the addition or organic polymers, and heat conditioning.

2. The two general ways that conditioned sludge can be dewatered are air drying and by some mechanical means.

3. The two basic steps that happen during chemical sludge conditioning are coagulation and flocculation.

4. The electrical charge that sludge particles usually have is negative.

5. The amount of sludge conditioning chemical increases as the size of the sludge particles gets smaller.

6. Two popular sludge conditioning chemicals are lime and ferric chloride.

7. When lime is mixed with water heat is generated.

8. Three advantages of using lime for sludge conditioning are pH control, odor reduction, and sludge disinfection.

9. Two disadvantages of using lime for sludge conditioning are an increase in sludge production of up to 30% and a lowered BTU content.

251

10. A polymer spill should be cleaned up by covering with kitty-litter type absorbent material and allowing it to soak up. Then remove by shovel.

11. The dry polymer's "percent activity" is a measure of the percentage of the contents that will form polymers.

12. Anionic polymers have a negative charge.

13. Cationic polymers are used most frequently in sludge conditioning because they have a positive charge which combines with the negatively charged sludge particles to cancel them out.

14. The range of polymer solution concentration used typically ranges from 0.1 to 1.0%.

15. As a polymer solution is aged the long polymer chains uncoil.

16. The chemical, when added upstream of the polymer, that can reduce the amount of polymer used is potassium permanganate.

17. The piece of kitchen equipment that is most like the thermal conditioning process is a pressure cooker.

18. The two actions that take place on a sludge drying bed are drainage and evaporation.

19. The type of climates in which sludge drying beds work best are warm and dry.

20. There are two different belts used in a belt filter press.

21. The four different areas in a belt filter press are the polymer addition point, the gravity drainage section, the "wedge zone," and the area with the rollers.

22. Two chemicals that might be added to the sludge to allow the dried solids to let go of the belts more easily are anionic and nonionic polymers.

23. In a solid bowl centrifuge, the two things that are spinning are the solid bowl and the scroll conveyor.

24. At minimum differential speed the sludge cake would be the dryest.

25. Pressure applied to the sludge in a belt filter press is between 200 and 250 psi.

26. In order to be effective, a vacuum filter usually requires the sludge to be thermally conditioned.

SOLUTIONS TO THE PROBLEMS: SLUDGE DEWATERING

1.
$$200 \text{ ft} \times 22 \text{ ft} \times 11 \text{ in} \times \frac{1 \text{ ft}}{12 \text{ in.}} \times \frac{7.48 \text{ gal}}{1 \text{ ft}^3} = 30,169.3 \text{ gal}$$

$$30,169.3 \text{ gal} \times \frac{3.1\%}{100\%} \times \frac{8.34 \text{ lb}}{1 \text{ gal}} = 7800 \text{ lb}$$

$$\frac{7800 \text{ lb}}{23 \text{ d}} = 339.13 \text{ lb/d}$$

$$\frac{339.13 \text{ lb}}{1 \text{ d}} \times \frac{365 \text{ d}}{1 \text{ yr}} = 123,782 \text{ lb/yr}$$

$$\frac{123,782 \text{ lb}}{1 \text{ yr}} \times \frac{1}{(200 \text{ ft} \times 22 \text{ ft})} = 28.1 \text{ lb/yr/ft}^2$$

2.
$$61 \text{ m} \times 6.7 \text{ m} \times 28 \text{ cm} \times \frac{1 \text{ m}}{100 \text{ cm}} \times \frac{1000 \text{ L}}{1 \text{ m}^3} = 114,436 \text{ L}$$

$$114,436 \text{ gal} \times \frac{3.1\%}{100\%} \times \frac{1 \text{ kg}}{1 \text{ L}} = 3547.5 \text{ kg}$$

$$\frac{3547.5 \text{ kg}}{23 \text{ d}} = 154.2 \text{ kg/day}$$

$$\frac{154.2 \text{ kg}}{1 \text{ d}} \times \frac{365 \text{ d}}{1 \text{ yr}} = 56,283 \text{ kg/yr}$$

$$\frac{56,283 \text{ kg}}{1 \text{ yr}} \times \frac{1}{(61 \text{ m} \times 6.7 \text{ m})} = 137.7 \text{ kg/yr/m}^2$$

3.
$$\frac{0.18 \text{ MG}}{1 \text{ d}} \times \frac{1,000,000 \text{ gal}}{1 \text{ MG}} \times \frac{1 \text{ d}}{1440 \text{ min}}$$

$$= 125 \text{ gal/min} = 125 \text{ gpm}$$

$$68 \text{ in.} \times \frac{1 \text{ ft}}{12 \text{ in.}} = 5.67 \text{ ft}$$

$$\frac{125 \text{ gpm}}{5.67 \text{ ft}} = 22.0 \text{ gpm/ft}$$

4. $$\frac{0.68 \text{ ML}}{1 \text{ d}} \times \frac{1,000,000 \text{ L}}{1 \text{ ML}} \times \frac{1 \text{ d}}{1440 \text{ min}} \times \frac{1 \text{ min}}{60 \text{ s}} = \frac{7.87 \text{ L}}{1 \text{ s}}$$

$$173 \text{ cm} \times \frac{1 \text{ m}}{100 \text{ cm}} = 1.73 \text{ m}$$

$$\frac{7.87 \text{ L/s}}{1.73 \text{ m}} = 4.55 \text{ L/s/m}$$

5. $$950 \text{ gal} \times \frac{4.0\%}{100\%} \times \frac{8.34 \text{ lb}}{1 \text{ gal}} = 316.92 \text{ lb}$$

$$\frac{316.92 \text{ lb}}{145 \text{ min}} \times \frac{60 \text{ min}}{1 \text{ hr}} = 131.14 \text{ lb/hr}$$

$$\frac{131.14 \text{ lb/hr}}{148 \text{ ft}^2} = 0.89 \text{ lb/hr/ft}^2$$

6. $$3600 \text{ L} \times \frac{4.0\%}{100\%} \times \frac{1 \text{ kg}}{1 \text{ L}} = 144 \text{ kg}$$

$$\frac{144 \text{ kg}}{145 \text{ min}} \times \frac{60 \text{ min}}{1 \text{ hr}} = 59.6 \text{ kg/hr}$$

$$\frac{59.6 \text{ kg/hr}}{13.8 \text{ m}^2} = 4.32 \text{ kg/hr/m}^2$$

7. $$\text{Area} = 3.14 \times 9.8 \text{ ft} \times 12 \text{ ft} = 369.3 \text{ ft}^2$$

$$\frac{35 \text{ gal}}{1 \text{ min}} \times \frac{60 \text{ min}}{1 \text{ hr}} \times \frac{8.34 \text{ lb}}{1 \text{ gal}} \times \frac{13\%}{100\%} = 2276.8 \text{ lb/hr}$$

$$\frac{2276.8 \text{ lb/hr}}{369.3 \text{ ft}^2} = 6.2 \text{ lb/hr/ft}^2$$

8. $$\text{Area} = 3.14 \times 3.00 \text{ m} \times 3.65 \text{ m} = 34.38 \text{ m}^2$$

$$\frac{2.21 \text{ L}}{1 \text{ s}} \times \frac{60 \text{ s}}{1 \text{ min}} \times \frac{60 \text{ min}}{1 \text{ hr}} \times \frac{1 \text{ kg}}{1 \text{ L}} \times \frac{13\%}{100\%} = 1034.3 \text{ kg/hr}$$

$$\frac{1034.3 \text{ kg/hr}}{34.38 \text{ m}^2} = 30.1 \text{ kg/hr/m}^2$$

9.
$$\text{Percent Recovery} = \frac{\text{lb/h Cake Solids}}{\text{lb/h Sludge Solids}} \times 100\%$$

$$\text{lb/h Cake Solids} = \frac{3070 \text{ lb}}{1 \text{ hr}} \times \frac{45\%}{100\%} = 1381.5 \text{ lb/h}$$

$$\text{lb/h Sludge Solids} = \frac{23 \text{ gal}}{1 \text{ min}} \times \frac{60 \text{ min}}{1 \text{ hr}} \times \frac{8.51 \text{ lb}}{1 \text{ gal}} \times \frac{12\%}{100\%}$$

$$= 1409.3 \text{ lb/h}$$

$$\% \text{ Recovery} = \frac{1381.5 \text{ lb/h}}{1409.3 \text{ lb/h}} \times 100\% = 98\%$$

10.
$$\text{Percent Recovery} = \frac{\text{kg/h Cake Solids}}{\text{kg/h Sludge Solids}} \times 100\%$$

$$\text{kg/h Cake Solids} = \frac{1392 \text{ kg}}{1 \text{ hr}} \times \frac{45\%}{100\%} = 626.4 \text{ kg/h}$$

$$\text{kg/h Sludge Solids} = \frac{1.45 \text{ L}}{1 \text{ s}} \times \frac{3600 \text{ sec}}{1 \text{ hr}} \times \frac{1.02 \text{ kg}}{1 \text{ L}} \times \frac{12\%}{100\%}$$

$$= 638.9 \text{ kg/h}$$

$$\% \text{ Recovery} = \frac{626.4 \text{ kg/h}}{638.9 \text{ kg/h}} \times 100\% = 98\%$$

ANSWERS TO THE QUESTIONS: ODOR CONTROL

1. The two basic categories of odors in a wastewater treatment plant are those from inorganic chemicals and those from organic chemicals.

2. The two chemicals responsible for the majority of the inorganic odors are hydrogen sulfide and ammonia.

3. Hydrogen sulfide is the one that smells like rotten eggs.

4. When hydrogen sulfide is dissolved in water, it can form the hydrosulfide ion ($-HS$).

5. No, this ion is not odorous.

6. Above a pH of 8.0, the hydrogen sulfide will stay in this ionic form.

7. The two chemicals that can be added to the wastewater to raise the pH are lime and sodium hydroxide.

8. The two different forms of ammonia are ammonia gas (NH_3) and the ammonium ion (NH_4^+).

9. At a pH below 9.0, the ammonia in water is mostly the non-odorous ammonium ion (NH_4^+).

10. Inside a pH range of 8 to 9 the odors of both hydrogen sulfide and ammonia will be minimized.

11.   There are thousands of different organic odors.

12.   The lowest concentration of an odor that the human nose can smell is 0.1 ppb or one tenth of a part per billion.

13.   Odorous air can be sampled by expanding a collapsible sample bottle in the area where the odor is strongest.

14.   If an odorous air sample is diluted with odor-free air to the point that at least half the people on an odor panel can't smell it, this concentration is considered to be the minimum concentration detected by the "average" person.

15.   GC stands for Gas Chromatography.

16.   The GC test tell us only which chemicals are in an odorous air sample.

17.   MS stands for Mass Spectrometry.

18.   The MS test tells us how much of a given chemical is present.

19.   The human nose can detect odors at a lower concentration than the best available equipment.

20.   The best place to treat odors is at the source.

21.   The two chemicals that are more commonly used in gravity sewers to treat odors associated with hydrogen sulfide are sodium hypochlorite and potassium permanganate.

22.   The chemical recommended for use in force mains is calcium hypochlorite.

23.   Another chemical that has been used to treat hydrogen sulfide is hydrogen peroxide.

24.   Hydrogen peroxide does not treat odors from ammonia or organic chemicals.

25.   Hydrogen sulfide will not form if the DO is kept above 1.9 mg/L.

26.   Sodium nitrate could be added to the wastewater to increase the available oxygen.

27.   Inorganic odors can be effectively treated by a scrubber.

28.   Scrubbers do not treat organic odors very well because the organic chemicals do not dissolve well in the water, where the chlorine or permanganate is.

29.   The two methods used to effectively treat organic odors are vapor combustion units (VCUs) and activated carbon.

30.   A masking agent is a pleasant smelling odorous chemical that is sprayed in an attempt to cover the smell of a bad odor.

SOLUTIONS TO THE PROBLEMS: ODOR CONTROL

1. a. $\dfrac{4.5\,\text{MG}}{1\,\text{d}} \times 0.2\,\text{mg/L} \times \dfrac{8.34\,\text{lb}}{1\,\text{gal}} \times \dfrac{10\,\text{lb NaNO}_3}{1\,\text{lb H}_2\text{S}} \times 7\,\text{d}$

$$= 525.4\,\text{lb} \times \dfrac{1\,\text{bag}}{40\,\text{pounds}} = 13.13\,\text{bags}$$

      14 bags would be needed.

  b. It would be better if the problem could be treated at its source. There are at least three potential sources:

      (1) An industrial discharge

      (2) A long interceptor trunk

      (3) A pumping station with an oversized wet well

2. a. $17\,\text{ML/d} \times 0.2\,\text{mg/L} \times \dfrac{10\,\text{kg NaNO}_3}{1\,\text{kg H}_2\text{S}} \times 7\,\text{d}$

$$= 238\,\text{kg} \times \dfrac{1\,\text{bag}}{18\,\text{kg}} = 13.2\,\text{bags}$$

      14 bags would be needed.

  b. It would be better if the problem could be treated at its source. There are at least three potential sources:

      (1) An industrial discharge

      (2) A long interceptor trunk

      (3) A pumping station with an oversized wet well

3. a. $$\frac{0.6\,\text{MG}}{1\,\text{d}} \times 20\,\text{mg/L} \times \frac{8.34\,\text{lb}}{1\,\text{gal}} = 100.08\,\text{lb}\,Cl_2/d$$

$$100.08\,\text{lbs}\,Cl_2/d \times \frac{100\%}{60\%} = 166.8\,\text{lbs calcium hypochlorite/d}$$

$$\frac{166.8\,\text{lb}}{1\,\text{d}} \times 30\,\text{d} = 5004\,\text{lb calcium hypochlorite for 30 d}$$

   b.   Sodium hypochlorite is more highly recommended for gravity sewers. Testing with sodium hypochlorite may prove it to be more effective on the reduction of odors.

4. a. $$\frac{2270\,\text{m}^3/d \times 20\,\text{mg/L}}{1000} = 45.4\,\text{kg}\,Cl_2/d$$

$$45.4\,\text{kg}\,Cl_2/d \times \frac{100\%}{60\%} = 75.7\,\text{kg calcium hypochlorite/d}$$

$$\frac{75.7\,\text{kg}}{1\,\text{d}} \times 30\,\text{d} = 2271\,\text{kg calcium hypochlorite for 30 d}$$

   b.   Sodium hypochlorite is more highly recommended for gravity sewers. Testing with sodium hypochlorite may prove it to be more effective on the reduction of odors.

5. a. $$\frac{1.0\,\text{MG}}{1\,\text{d}} \times 12\,\text{mg/L} \times \frac{8.34\,\text{lb}}{1\,\text{gal}} \times \frac{2.5\,\text{lbs}\,H_2O_2}{1\,\text{lb}\,H_2S}$$

$$= 250.2\,\text{lbs}\,H_2O_2$$

$$\frac{250\,\text{lb}\,H_2O_2}{1\,\text{d}} \times \frac{100\%}{75\%} \times \frac{1\,\text{gal}}{8.34\,\text{lb}} = 40\,\text{gal/d}$$

   b.   The treatment will be effective to treat the hydrogen sulfide odors. However, it will do nothing for the ammonia odors. A form of chlorine would be required to take care of both. Sodium hypochlorite would probably be best.

6. a. $$\frac{3785\,\text{m}^3/d \times 12\,\text{mg/L}}{1000} \times \frac{2.5\,\text{kg}\,H_2O_2}{1\,\text{kg}\,H_2S} = 113.5\,\text{kg}\,H_2O_2$$

$$\frac{113.5 \text{ kg H}_2\text{O}_2}{1 \text{ d}} \times \frac{100\%}{75\%} \times \frac{1 \text{ L}}{1 \text{ kg}} = 151 \text{ L/d}$$

b.   The treatment will be effective to treat the hydrogen sulfide odors. However, it will do nothing for the ammonia odors. A form of chlorine would be required to take care of both. Sodium hypochlorite would probably be best.

7 a.
$$450 \text{ ft} \times 670 \text{ ft} \times \frac{1 \text{ ac}}{43,560 \text{ ft}^2} = 6.92 \text{ ac}$$

$$\left(6.92 \text{ ac} \times \frac{100 \text{ lb}}{1 \text{ ac}}\right) + 6 \times \left(6.92 \text{ ac} \times \frac{50 \text{ lb}}{1 \text{ ac}}\right) = 2768 \text{ lb}$$

b.   Oxygen could be added by using floating aerators or by driving across the surface in a motor boat. Some of the pond's effluent could be recirculated to dilute the incoming waste strength. Look for specific sources of septic or high strength influent that could be eliminated or pretreated.

8. a.
$$137 \text{ m} \times 204 \text{ m} \times \frac{1 \text{ ha}}{10,000 \text{ m}^2} = 2.79 \text{ ha}$$

$$\left(2.79 \text{ ha} \times \frac{112 \text{ kg}}{1 \text{ ha}}\right) + 6 \times \left(2.79 \text{ ha} \times \frac{56 \text{ kg}}{1 \text{ ha}}\right) = 1250 \text{ kg}$$

b. Oxygen could be added by using floating aerators or by driving across the surface in a motor boat. Some of the pond's effluent could be recirculated to dilute the incoming waste strength. Look for specific sources of septic or high strength influent that could be eliminated or pretreated.

ANSWERS TO THE QUESTIONS: LAB

1. The pH is a measure of how much acid is in the water.

2. One way to describe an acid is by its ability to make hydrogen ions in water solution.

3. An example of a chemical that is a base is sodium hydroxide (NaOH).

4. If an acid solution is mixed with a base solution so that they neutralize each other, what is left in the solution is a salt and water.

5. A neutral solution has a pH value of 7.0.

6. The concentration of hydrogen ions at a pH of 7.0 is 0.1 μg/L.

7. The concentration of hydrogen ions at a pH of 6.0 is 1.0 μg/L.

8. The concentration of hydrogen ions at a pH of 5.0 is 10 μg/L.

9. As the pH goes down by 1.0 units, the concentration of hydrogen ions becomes 10 times greater.

10. The chemical used to supply oxygen in the COD test is potassium dichromate.

11. The BOD is between 50% and 70% as large as the COD.

12. The biggest advantage of the COD over the BOD is the time to do the test, three hours versus five days.

13. The most common form of phosphorus in wastewaters is orthophosphate (H_3PO_4).

14. During the first step of a test for total phosphorus concentration, all forms of phosphorus are converted to the orthophosphate form.

15. After the samples being tested for phosphorus turn blue, a spectrophotometer is used to find the exact phosphorus concentration.

16. The bug used most frequently as an indicator organism is fecal coliform.

17. The chemical used most often to dechlorinate an effluent sample before doing biological testing is sodium thiosulfate.

18. The test used most often to count the indicator bugs in the plant effluent is the membrane filter test.

19. A dense population of bugs grown on one spot where only one bug had started is called a colony.

20. In a membrane filter test for fecal coliform there are black, blue, and yellow dots. Only the blue ones should be counted.

21. When a membrane filter test for fecal coliform is run, there is more than one sample volume used. All results between a count of 20 and 60 are used in the calculation.

22. The results of a membrane filter test are reported as count per 100 mL.

23. A seeded BOD test is needed for some samples because either the bugs have been killed or there weren't any. In either case, bugs need to be added to the sample.

24. For a chlorinated effluent sample, the best source of "seed" bugs is in the settled sewage or primary effluent.

25. The name given to the final DO in a BOD test is the residual.

26. The name given to the difference between the initial and final DOs in a BOD test is the depletion.

27. The name given to the amount of oxygen used during a BOD test is also the depletion.

28. The depletion for the seed-alone sample represents the amount of oxygen the seed bugs used when they only had "seed" food to eat.

29. The depletion for the seeded sample represents the amount of oxygen the seed bugs used eating both effluent food and "seed" food.

30. If you found the difference between the depletions described in questions 28 and 29, this would represent the amount of oxygen the seed bugs used to eat just the effluent food.

31. The amount of oxygen used by the bugs in the seeded sample to eat only seed food is not measured directly. It is estimated based on the amount of oxygen they used in the seed-alone sample.

32. A "blank" in the BOD test is a BOD bottle with only dilution water that is incubated to see if the dilution water has any contamination.

33. Ammonia is added to a BOD test bottle in the dilution water as a nutrient for the carbonaceous bugs.

34. Yes, if nitrifiers are in a sample, they will eat the ammonia added to a BOD bottle.

35. If nitrifiers are in a sample, the BOD test is, therefore, not accurate.

36. If a chemical is added to the BOD bottle to prevent the nitrifiers from growing, the test is called a carbonaceous BOD or CBOD.

37. The NOD is the nitrogenous oxygen demand.

38. Two ways lab chemical concentrations may be given are in terms of molarity and normality.

39. Molarity is concentration as moles per liter.

40. A burette is a clear tube with markings on the side and a valve on the bottom.

41. GAW stands for Gram Atomic Weight.

42. The equivalent weight of an acid is the weight of an acid needed to neutralize one mole of hydroxide ions (OH^-).

43. The equivalent weight of a base is the weight of a base needed to neutralize one mole of hydrogen ions (H^+).

44. Normality is a chemical solution concentration expressed in terms of equivalents per liter.

SOLUTIONS TO THE PROBLEMS: LAB

1. In this case two of the counts fall into the desired range so both are used. In the calculation the counts are added together and the sample volumes are added together.

$$\frac{count}{100 \text{ mL}} = \frac{(51 + 25)}{(10 \text{ mL} + 5 \text{ mL})} \times \frac{100 \text{ mL}}{100 \text{ mL}}$$

$$= \frac{76 \times 100}{15} \times \frac{1}{100 \text{ mL}}$$

$$= 507/100 \text{ mL}$$

2. The counts obtained were:

Volume:	50 mL	25 mL	10 mL
Count:	18	8	<1

When all three counts fall below 20, then all three are used.

$$\frac{count}{100 \text{ mL}} = \frac{(18 + 8 + 0)}{(50 \text{ mL} + 25 \text{ mL} + 10 \text{ mL})} \times \frac{100 \text{ mL}}{100 \text{ mL}}$$

$$= \frac{26 \times 100}{85} \times \frac{1}{100 \text{ mL}}$$

$$= 31/100 \text{ mL}$$

3.
$$BOD = \frac{DO_i - DO_f}{\text{Sample fraction}} = \frac{6.94 \text{ mg/L} - 4.36 \text{ mg/L}}{0.05}$$

$$= \frac{2.58 \text{ mg/L}}{0.05} = 51.6 \rightarrow 52 \text{ mg/L}$$

4.
$$BOD = (DO_i - DO_f) \times \frac{300 \text{ mL}}{\text{Sample mL}}$$

Initial DO $= (7.96 \text{ mg/L} \times 0.876) + (2.43 \text{ mg/L} \times 0.12)$

$+ (0.0 \text{ mg/L} \times 0.004)$

$= 7.26 \text{ mg/L}$

DO_i of the seed alone sample

Dilution water fraction $= 1 - 0.05$

$= 0.95$

Initial DO $= (7.96 \text{ mg/L} \times 0.95) + (0.0 \text{ mg/L} \times 0.004)$

$= 7.56 \text{ mg/L}$

$$\text{Depletion} = \underset{\text{(Seeded Eff)}}{(DO_i - DO_f)} - \left[\underset{\text{(Seed Alone)}}{(DO_i - DO_f)} \right.$$

$$\left. \times \frac{(\% \text{ Seed in Effluent})}{(\% \text{ Seed Alone})} \right]$$

$= (7.26 \text{ mg/L} - 3.98 \text{ mg/L}) - (7.56 \text{ mg/L} - 4.12 \text{ mg/L})$

$\times \dfrac{(0.4\%)}{(5\%)}$

$= 3.00 \text{ mg/L}$

$$\text{BOD} = \frac{\text{Depletion}}{\text{Sample Fraction}} = \frac{3.00 \text{ mg/L}}{0.12} = 25 \text{ mg/L}$$

11. $\text{Normality(1)} \times \text{Volume(1)} = \text{Normality(2)} \times \text{Volume(2)}$

$0.05 \text{ N} \times 2000 \text{ mL} = 10 \text{ N} \times \text{Volume(2)}$

$\dfrac{0.05 \text{ N}}{10 \text{ N}} \times 2000 \text{ mL} = \dfrac{10 \text{ N}}{10 \text{ N}} \times \text{Volume (2)}$

$\dfrac{0.05 \text{ N}}{10 \text{ n}} \times 2000 \text{ mL} = \text{Volume (2)}$

$= 10 \text{ mL}$

$= 7.43 \text{ mg/L} - 4.21 \text{ mg/L}) \times \dfrac{300 \text{ mL}}{10 \text{ mL}}$

$= 96.6 \rightarrow 97 \text{ mg/L}$

---

5. $$\text{Depletion} = \underset{\text{(Seeded Eff)}}{(DO_i - DO_f)} - \left[\underset{\text{(Seed Alone)}}{(DO_i - DO_f)} \right.$$

$$\left. \times \frac{(\% \text{ Seed in Effluent})}{(\% \text{ Seed Alone})} \right]$$

$= (7.29 \text{ mg/L} - 4.62 \text{ mg/L})$

$- (7.58 \text{ mg/L} - 4.89 \text{ mg/L}) \times \dfrac{(0.6\%)}{(4\%)}$

$= 2.27 \text{ mg/L}$

$$\text{BOD} = \frac{\text{Depletion}}{\text{Sample Fraction}} = \frac{2.27 \text{ mg/L}}{0.18} = 12.6 \rightarrow 13 \text{ mg/L}$$

---

6. $$\text{Depletion} = \underset{\text{(Seeded Eff)}}{(DO_i - DO_f)} - \left[\underset{\text{(Seed Alone)}}{(DO_i - DO_f)} \right.$$

$$\left. \times \frac{(\% \text{ Seed in Effluent})}{(\% \text{ Seed Alone})} \right]$$

$= (6.86 \text{ mg/L} - 3.95 \text{ mg/L})$

$- (7.44 \text{ mg/L} - 4.62 \text{ mg/l}) \times \dfrac{(0.5\%)}{(5\%)}$

$= 2.63 \text{ mg/L}$

$$\text{BOD} = \frac{\text{Depletion}}{\text{Sample Fraction}} = \frac{2.63 \text{ mg/L}}{0.25} = 10.52 \rightarrow 11 \text{ mg/L}$$

---

7. DO_i of the seeded effluent sample

Dilution water fraction $= 1 - (0.12 + 0.01)$

$= 1 - 0.13$

$$= 0.87$$

$$\text{Initial DO} = (8.1 \text{ mg/L} \times 0.87) + (5.2 \text{ mg/L} \times 0.12)$$

$$+ (2.0 \text{ mg/L} \times 0.01)$$

$$= 7.69 \text{ mg/L}$$

DO_i of the seed alone sample

$$\text{Dilution water fraction} = 1 - 0.06$$

$$= 0.94$$

$$\text{Initial DO} = (8.1 \text{ mg/L} \times 0.94) + (2.0 \text{ mg/L} \times 0.06)$$

$$= 7.73 \text{ mg/L}$$

---

8. DO_i of the seeded effluent sample

$$\text{Dilution water fraction} = 1 - (0.18 + 0.006)$$

$$= 1 - 0.186$$

$$= 0.814$$

$$\text{Initial DO} = (7.9 \text{ mg/L} \times 0.814) + (4.8 \text{ mg/L} \times 0.18)$$

$$+ (0.0 \text{ mg/L} = 0.006)$$

$$= 7.29 \text{ mg/L}$$

DO_i of the seed alone sample

$$\text{Dilution water fraction} = 1 - 0.04$$

$$= 0.96$$

$$\text{Initial DO} = (7.9 \text{ mg/L} \times 0.96) + (0.0 \times 0.04)$$

$$= 7.58 \text{ mg/ L}$$

---

9. DO_i of the seeded effluent sample

$$\text{Dilution water fraction} = 1 - (0.18 + 0.01)$$

$$= 1 - 0.19$$

$$= 0.81$$

$$\text{Initial DO} = (8.1 \times 0.81) + (4.6 \times 0.18) + (1.5 \times 0.01)$$

$$= 7.40 \text{ mg/L}$$

DO_i of the seed alone sample

$$\text{Dilution water fraction} = 1 - 0.08$$

$$= 0.92$$

$$\text{Initial DO} = (8.1 \times 0.92) + (1.5 \times 0.08)$$

$$= 7.57 \text{ mg/L}$$

$$\text{Depletion} = \underset{\text{(Seeded Eff)}}{(DO_i - DO_f)} - \left[\underset{\text{(See Alone)}}{(DO_i - DO_f)} \times \frac{(\% \text{ Seed in Effluent})}{(\% \text{ Seed Alone})}\right]$$

$$= (7.40 \text{ mg/L} - 4.23 \text{ mg/L}) - (7.57 \text{ mg/L} - 4.92 \text{ mg/L})$$

$$\times \frac{(1\%)}{(8\%)}$$

$$= 2.84 \text{ mg/L}$$

$$\text{BOD} = \frac{\text{Depletion}}{\text{Sample Fraction}} = \frac{2.84 \text{ mg/L}}{0.18} = 15.78 = \text{ >}$$

---

10. DO_i of the seeded effluent sample

$$\text{Dilution water fraction} = 1 - (0.12 + 0.004)$$

$$= 1 - 0.124$$

$$= 0.876$$

12.   Normality(1) × Volume(1) = Normality(2) × Volume(2)

5 N × 10 mL = Normality(2) × 8.4 mL

$$5 \text{ N} \times \frac{10 \text{ mL}}{8.4 \text{ mL}} = \text{Normality(2)}$$

$$\frac{5 \text{ N} \times 10 \text{ mL}}{8.4 \text{ mL}} = \text{Normality(2)}$$

$$= 5.95 \text{ N}$$